501
Critical Reading Questions

501

Critical Reading Questions

LEARNINGEXPRESS ®

NEW YORK

Library of Congress Cataloging-in-Publication Data:
    501 critical reading questions.—1st ed.
        p. cm.
    ISBN 978-1-57685-510-2 (pbk. : alk. paper)
    1. Reading (Secondary)—Examinations, questions, etc. 2. Reading comprehension—
Examinations, questions, etc. 3. Readers (Secondary) I. Title: Five hundred one
critical reading questions. II. Title: Five hundred and one critical reading questions.
III. LearningExpress (Organization)
    LB1632.A16 2004
    428.4'07'12—dc22

                                                                    2004001114

Printed in the United States of America

9 8

First Edition

ISBN 978-1-57685-510-2

For more information or to place an order, contact LearningExpress at:
    2 Rector Street
    26th Floor
    New York, NY 10006

Or visit us at:
    www.learnatest.com

The LearningExpress Skill Builder in Focus Writing Team is comprised of experts in test preparation, as well as educators and teachers who specialize in language arts.

LearningExpress Skill Builder in Focus Writing Team

Marco A. Annunziata
Freelance Writer
New York, New York

Elizabeth Chesla
English Instructor
Language Arts Expert
Harleysville, Pennsylvania

Brigit Dermott
Freelance Writer
English Tutor, New York Cares
New York, New York

Margaret Muirhead
Freelance Writer
Arlington, Massachusetts

Patricia Mulrane
Freelance Writer
New York, New York

Lauren Starkey
Freelance Reference Writer
Essex, Vermont

C Reed
Test Preparation Expert
Burbank, California

Contents

Introduction

Why Should I Use this Book?

Schools and employers know that students and workers who reason criti-
cally about what they read are better students and more valuable employees. That is why standardized tests almost invariably include a reading
comprehension section.

This book is designed to help you be a more successful critical reader.
You are probably most interested in performing well on a standardized test
such as the SAT, ACT, or a vocational or professional exam. By reading and
working through *501 Critical Reading Questions* you will become much more
proficient at answering the multiple-choice questions found on those tests.
The benefits you gain from this practice and from your conscious attention
to critical reasoning skills will extend far beyond any exam and into all
aspects of your life. Reading will become a much more rewarding and
enjoyable experience, and your life will be richer for it.

What Is in this Book?

Each of the chapters in this book focuses on a different subject matter, so
regardless of the exact exam you need to prepare for, there will be content
similar to material you will face on your exam. However, it's important that

you practice with all the passages, not just the ones in your areas of interest. Sometimes unfamiliar subjects can teach you the most valuable lessons about critical reading.

Each chapter contains three short reading passages, similar to the ones found on many exams, including the SAT. There are also six longer passages, two of which are paired for purposes of comparison.

Passages in Chapter One deal with popular culture and current events. History and politics are covered in Chapter Two. Chapter Three's passages focus on the humanities—they are drawn from fields such as mythology, philosophy, and the arts. Chapter Four has passages that deal with health and medicine. Chapter Five draws passages from literature. Chapter Six's passages are drawn from the field of music. Chapter Seven contains material on science and nature. Chapter Eight covers sports and leisure. And finally, Chapter Nine's passages are based in the social sciences of anthropology and sociology.

STAY ACTIVE

The most important thing to know about critical reading is that it is an active endeavor. Keep your mind active and on its figurative toes at all times. Underline important points as you read, argue with the author, make notes, and do whatever you need to stay involved with the passage.

Seven Strategies for Success

Even though short passages are new to the SAT, strategies for successfully answering the questions are identical to those for the longer passages. The first thing you will want to do, before diving into the practice, is to make sure you are thoroughly familiar with these strategies. Then feel free to adapt them to suit your needs and preferences. One word of caution, though: Be sure you actually try each strategy several times before deciding whether or not it suits you!

1. *Get involved with the passage.* Critical reading is an active endeavor, not a passive one. React to the material, form questions as you read, and make your own marks on the paper. Write in the margins, underline important words and sentences—talk back!

2. *Try looking at the questions* (but not the answers) *before you read the passage.* Make sure you understand what each question is asking. What are the key words in the questions? Are there phrases you can look for in the passage? If so, underline them or jot them in the margin so that you can look for them in the passage. Then, when you find them, you can either answer the question right away or mark the area to return to later.

3. *After reading the passage, return to the questions and try to answer each one in your own words before you look at your answer choices.* The reason for this is that the answers will contain distracter choices. These are choices that are logically plausible but not correct, that contain words and phrases found in the passage but are not correct, or that are close to correct but wrong in some detail. If you can formulate your own answer before looking at your choices, you are less likely to be lured by an incorrect answer choice.

4. *As with all multiple-choice questions, elimination is an important strategy when you aren't sure of the answer.* Usually you can narrow down your choices to two or three without too much effort. When you eliminate an incorrect choice, it's important to actually cross it out in your test booklet so that you aren't distracted by it again as you focus on the remaining possibilities.

5. *Refer back to the passage(s) on virtually every question.* Even if you think you know the answer to a question without looking at the passage, look anyway, just to confirm your answer and to make sure you haven't fallen for a clever distracter.

6. *When you encounter a two-passage section, read the passages with their relationship in mind.* Are they opposed or in agreement? If there is some other type of relationship, how would you describe it? If the passages have opposing viewpoints, what are the points of difference? You may want to make notes about these things in the margin.

7. *Don't be afraid to skip around among the questions, or among the passages within a section.* This is an especially important strategy if you know from past experience that you often run out of time on standardized tests. If this is the case, and you encounter a passage you're having difficulty with, go on to the next one and come back to the difficult one later, as time allows.

Remind Me Why I'm Doing This

Finally, as you work through these 501 questions, think of it as time spent doing something for yourself. It is extremely important for you to improve your critical reading skills, not only for standardized tests, but also for your success throughout life. And, besides, there is some pretty interesting stuff in this book! Enjoy.

501
Critical Reading Questions

Popular Culture

Questions 1–3 are based on the following passage.

The following selection is about the invention of the compact disc, and explains how it works.

(1)   Compact discs (CDs), which may be found in over 25 million American homes, not to mention backpacks and automobiles, first entered popular culture in the 1980s. But their history goes back to the 1960s, when an inventor named James Russell decided to create an alterna-

(5)   tive to his scratched and warped phonograph records—a system that could record, store, and replay music without ever wearing out.

    The result was the compact disc (CD). Made from 1.2 mm of polycarbonate plastic, the disc is coated with a much thinner aluminum layer that is then protected with a film of lacquer. The lacquer layer

(10)   can be printed with a label. CDs are typically 120 mm in diameter, and can store about 74 minutes of music. There are also discs that can store 80, 90, 99, and 100 minutes of music, but they are not as compatible with various stereos and computers as the 74–minute size.

    The information on a standard CD is contained on the polycar-

(15)   bonate layer, as a single spiral track of pits, starting at the inside of the disk and circling its way to the outside. This information is read by shining light from a 780 nm wavelength semiconductor laser through the bottom of the polycarbonate layer. The light from the laser follows

1

(20) the spiral track of pits, and is then reflected off either the pit or the aluminum layer. Because the CD is read through the bottom of the disc, each pit looks like a bump to the laser.

Information is read as the laser moves over the bumps (where no light will be reflected) and the areas that have no bumps, also known as land (where the laser light will be reflected off the aluminum). The (25) changes in reflectivity are interpreted by a part of the compact disc player known as the detector. It is the job of the detector to convert the information collected by the laser into the music that was originally recorded onto the disc. This invention brought 22 patents to James Russell, who today says he working on an even better system for (30) recording and playing back music.

1. According to the passage, why did James Russell invent the CD?
   a. He was tired of turning over his records to hear both sides.
   b. He wanted to record more music on a new format.
   c. He wanted a purer, more durable sound than he could get from vinyl records.
   d. He was interested in getting patents.
   e. He wanted to work with lasers.

2. What would happen if the detector on a CD player malfunctioned?
   a. The spiral track would not be read properly.
   b. The pits and land would look like one unit.
   c. The changes in reflectivity would be absorbed back into the laser.
   d. The music would play backwards.
   e. The information read by the laser would not be converted into music.

3. Paragraph 3, lines 14–21, explains all of the following EXCEPT
   a. how the information on a CD is read.
   b. why semiconductor lasers were invented.
   c. where information is stored on a CD.
   d. what pits and bumps are.
   e. the purpose of the aluminum layer of a CD.

Questions 4–6 are based on the following passage.

The selection that follows is about the current state of the modeling industry.

(1)   The beginning of the twenty-first century has been called the end of the supermodel era by fashion magazines, trend watchers, and news organizations around the world. The models are being replaced, so the theory goes, with actors. Check the covers of fashion magazines, and you

(5)   will find that many on any given month feature an actor, rather than a model. But, as with most trends, this is nothing new.

From its beginnings in the 1920s, the modeling industry has provided beautiful people to help sell everything from magazines to computers to vacation destinations. John Robert Powers, who opened the

(10)   first modeling agency in 1923, was a former actor who hired his actor friends to model for magazine advertisements. Cary Grant, Lucille Ball, and Princess Grace of Monaco were clients. However, for many models simply being "great-looking" was where their resumés began and ended. The height of popularity for them was in the 1980s and

(15)   1990s, the era of the supermodel. A handful of "perfect" women commanded salaries of up to $25,000 a day to walk catwalks at fashion shows, appear in print ads, and pose their way through commercials. They were celebrities, treated with all of the lavish attention usually paid to heads of state or rock stars.

(20)   But that was in the supermodel heyday. As designers and magazine editors began to favor more exotic and more "real" looking models, the modeling handful grew into an army. The demand for the perfect-looking select few dropped, and women who had quirky smiles, a few extra pounds, spiky hair, or were past their twenties, gained favor. This

(25)   group was joined by those who achieved success in some other venue, such as music (think Renee Fleming raving about a watch), sports (Tiger Woods happily devouring his Wheaties®), and acting (Danny Glover waxing rhapsodic over MCI). Iconic fashion designer Calvin Klein summed it up: "I don't think that people are that interested in

(30)   models anymore. It's not a great moment for the modeling industry. It says a lot about our society and I think it's good."

4. According to the passage, the author believes that
   a. today's fashion models are not as perfect looking as were the supermodels.
   b. people still respond to perfection in advertising.
   c. today's fashion models are thinner than those in the past.
   d. to be a model, one must be taller than average.
   e. in the 1980s, models were paid more than they are today.

5. The phrase in lines 13 and 14, *"great-looking" was where their resumes began and ended*, is
   a. a description of the models' work experience.
   b. meant to be taken literally.
   c. meant to be taken figuratively.
   d. a truthful statement.
   e. an example of pathos.

6. *Waxing rhapsodic* (line 28) most nearly means
   a. orchestrating a positive statement.
   b. becoming musical.
   c. burning a candle for.
   d. making overtures.
   e. becoming enthusiastic.

Questions 7–9 are based on the following passage.

This selection introduces the Computer Museum of America, and details an important item in its collection.

(1)   Wondering what to do with that old Atari Home Video Game in the attic? It's on the wish list of the Computer Museum of America, in San Diego, California, which hopes you will donate it to their holdings. The Museum was founded in 1983 to amass and preserve historic
(5)   computer equipment such as calculators, card punches, and typewriters, and now owns one of the world's largest collections. In addition, it has archives of computer-related magazines, manuals, and books that are available to students, authors, researchers, and others for historical research.
(10)   One item currently on display is a 1920s comptometer, advertised as "The Machine Gun of the Office." The comptometer was first sneered at by accountants and bookkeepers, many of whom could add four columns of numbers in their heads. The new machine was the first that could do the work faster than humans. The comptometer
(15)   gained a large following, and its operation became a formal profession that required serious training. But by the 1970s, computers took over, and comptometers, and the job of operating them, became obsolete.

7. All of the following are probably part of the collection of the
   Computer Museum of America EXCEPT
   a. adding machines.
   b. old computers.
   c. operation manuals for calculators.
   d. card punch machines.
   e. kitchen scales.

8. In line 12, the author used the words *sneered at* to show
   a. a negative image of accountants.
   b. what accountants and bookkeepers looked like.
   c. the negative reaction to the comptometer.
   d. the precursor of the comptometer operator.
   e. how fast accountants and bookkeepers could add.

9. What term paper topic could probably be researched at the
   Computer Museum of America?
   a. Alexander Graham Bell's contributions to American society
   b. IBM's contribution to the development of the modern
      computer
   c. more than just paintings: the museums of California
   d. the rise and fall of the comptometer operator
   e. why video games are harmful to our nation's youth

Questions 10–17 are based on the following passage.

*The following selection explains the origins and development of the modern
shopping mall.*

(1)   Today's shopping mall has as its antecedents historical marketplaces,
      such as Greek *agoras*, European *piazzas*, and Asian *bazaars*. The pur-
      pose of these sites, as with the shopping mall, is both economic and
      social. People go not only to buy and sell wares, but also to be seen,
(5)   catch up on news, and be part of the human drama. Both the market-
      place and its descendant the mall might also contain restaurants,
      banks, theaters, and professional offices.
         The mall is also the product of the creation of suburbs. Although
      villages outside of cities have existed since antiquity, it was the tech-
(10) nological and transportation advances of the 19th century that gave
      rise to a conscious exodus of the population away from crowded,
      industrialized cities toward quieter, more rural towns. Since the sub-
      urbs typically have no centralized marketplace, shopping centers or

malls were designed to fill the needs of the changing community, pro-
(15)   viding retail stores and services to an increasing suburban population.

The shopping mall differs from its ancient counterparts in a number of important ways. While *piazzas* and *bazaars* were open-air venues, the modern mall is usually enclosed. Since the suburbs are spread
out geographically, shoppers drive to the mall, which means that park-
(20)   ing areas must be an integral part of a mall's design. Ancient marketplaces were often set up in public spaces, but shopping malls are
designed, built, and maintained by a separate management firm as a
unit. The first shopping mall was built by J. C. Nichols in 1922 near
Kansas City, Missouri. The Country Club Plaza was designed to be an
(25)   automobile-centered plaza, as its patrons drove their own cars to it,
rather than take mass transportation as was often the case for city
shoppers. It was constructed according to a unified plan, rather than
as a random group of stores. Nichols' company owned and operated
the mall, leasing space to a variety of tenants.

(30)   The first enclosed mall was the Galleria Vittoria Emanuele in Milan,
Italy in 1865–77. Inspired by its design, Victor Gruen took the shopping
and dining experience of the Galleria to a new level when he created the
Southdale Center Mall in 1956. Located in a suburb of Minneapolis, it
was intended to be a substitute for the traditional city center. The 95-
(35)   acre, two-level structure had a constant climate-controlled temperature
of 72 degrees, and included shops, restaurants, a school, a post office,
and a skating rink. Works of art, decorative lighting, fountains, tropical
plants, and flowers were placed throughout the mall. Southdale afforded
people the opportunity to experience the pleasures of urban life while
(40)   protected from the harsh Minnesota weather.

In the 1980s, giant megamalls were developed. While Canada has
had the distinction of being home to the largest of the megamalls for
over twenty years, that honor will soon go to Dubai, where the Mall
of Arabia is being completed at a cost of over five billion U.S. dollars.
(45)   The 5.3 million square foot West Edmonton Mall in Alberta, Canada,
opened in 1981, with over 800 stores, 110 eating establishments, a
hotel, an amusement park, a miniature-golf course, a church, a zoo,
and a 438-foot-long lake. Often referred to as the "eighth wonder of
the world," the West Edmonton Mall is the number-one tourist
(50)   attraction in the area, and will soon be expanded to include more retail
space, including a facility for sports, trade shows, and conventions.

The largest enclosed megamall in the United States is Blooming-
ton, Minneapolis's Mall of America, which employs over 12,000 people. It has over five hundred retail stores, an amusement park which
(55)   includes an indoor roller coaster, a walk-through aquarium, a college,

and a wedding chapel. The mall contributes over one billion dollars each year to the economy of the state of Minnesota. Its owners have proposed numerous expansion projects, but have been hampered by safety concerns due to the mall's proximity to an airport.

10. The statement that people went to marketplaces to *be part of the human drama* (line 5) suggests that people
    a. prefer to shop anonymously.
    b. like to act on stage rather than shop.
    c. seem to be more emotional in groups.
    d. like to be in community, interacting with one another.
    e. prefer to be entertained rather than shop for necessities.

11. In line 1, *antecedents* most nearly means
    a. designers.
    b. planners.
    c. predecessors.
    d. role models.
    e. teachers.

12. All of the following questions can be explicitly answered on the basis of the passage EXCEPT
    a. Who designed the Southdale Center Mall in Minnesota?
    b. Why was the Country Club Plaza automobile-centered?
    c. What are three examples of historical marketplaces?
    d. Where is the Galleria Vittoria Emanuele?
    e. What is the Edmonton Mall often referred to as?

13. How was the Country Club Plaza different from an urban shopping district?
    a. It consisted of many more stores.
    b. It was built by one company that leased space and oversaw operations.
    c. It was enclosed.
    d. It had both retail stores and restaurants, and offered areas for community programs.
    e. It was based on an Italian design.

14. According to the passage, how did Southdale expand the notion of the shopping mall?

    a. It added an amusement park.

    b. It was unheated.

    c. It was the first to rise above two stories.

    d. It was designed with more parking spaces than any previous shopping mall.

    e. It was intended to be a substitute for the traditional city center.

15. According to paragraph 5, which is the only activity visitors to the West Edmonton Mall cannot enjoy?

    a. staying in a hotel

    b. gambling in a casino

    c. visiting animals in a zoo

    d. playing miniature golf

    e. riding an amusement park ride

16. When the author states in lines 38 and 39 that *Southdale afforded people the opportunity to experience the pleasures of urban life* she means that

    a. they could perform necessary and leisurely activities in one location.

    b. they could have a greater variety of retailers to choose from.

    c. they could see more artwork and botanicals than they would in a city.

    d. they could be entertained as they would be in a city.

    e. they could have taller buildings in their landscape.

17. What is NOT a probable reason for the proposed expansion of the Mall of America?

    a. so it can contribute more to the economy of its state

    b. to keep it closer in size to the other megamalls

    c. so it can employ more people

    d. to attract more tourists

    e. to compete for visitors with the Mall of Arabia

Questions 18–25 are based on the following passage.

The following selection explains the origins of sushi, and its popularity in the United States.

(1)   Burgers, fries, pizza, raw fish. Raw fish? Fast food in America is changing. *Sushi*, the thousand year old Japanese delicacy, was once thought of in this country as unpalatable and too exotic. But tastes have changed, for a number of reasons. Beginning in the 1970s, Americans

(5)   became increasingly more aware of diet and health issues, and began rejecting their traditional red-meat diets in favor of healthier, lower-fat choices such as fish, poultry, whole grains, rice, and vegetables. The way food was prepared began to change, too; rather than frying food, people started opting for broiled, steamed, and raw versions. *Sushi*, a

(10)   combination of rice and fish, fit the bill. In addition, that same decade saw Japan become an important global economic force, and companies began flocking to the country to do business. All things Japanese, including décor, clothing, and cuisine, became popular.

   Sushi started small in the United States, in a handful of restaurants

(15)   in big cities. But it caught on. Today, *sushi* consumption in American restaurants is 40% greater than it was in the late 1990s, according to the National Restaurant Association. The concession stands at almost every major league stadium sell *sushi*, and many colleges and universities offer it in their dining halls. But we're not just eating it out. The

(20)   National Sushi Association reports that there are over 5,000 *sushi* bars in supermarkets, and that number is growing monthly. This incredible growth in availability and consumption points to the fact that Americans have decided that *sushi* isn't just good for them, or just convenient, but that this once-scorned food is truly delicious.

(25)   The origins of this food trend may be found in Asia, where it was developed as a way of preserving fish. Fresh, cleaned fish was pressed between rice and salt and weighted with a heavy stone over a period of several months. During this time, the rice fermented, producing lactic acid that pickled and preserved the fish. For many years, the fish was

(30)   eaten and the rice was discarded. But about 500 years ago, that changed, and *hako-zushi* (boxed *sushi*) was created. In this type of *sushi*, the rice and fish are pressed together in a box, and are consumed together.

   In 1824, Yohei Hanaya of Edo (now called Tokyo) eliminated the fermentation process, and began serving fresh slices of seafood on

(35)   bases of vinegared rice. The vinegar was probably used to mimic the taste of fermented *sushi*. In fact, the word *sushi* actually refers to any vinegared rice dish, and not to the fish, as many Americans believe (the

fish is called *sashimi*). In Japanese, when *sushi* is combined with a modifier, it changes to the word *zushi*.
(40)   Chef Yohei's invention, called *nigiri zushi*, is still served today. It
now refers to a slice of fish (cooked or uncooked) that is pressed by
hand onto a serving of rice. Popular choices include *ama ebi* (raw
shrimp), *shime saba* (marinated mackerel), and *maguro* (tuna). In addition to the vinegar flavor in the rice, *nigiri zushi* typically contains a
(45) taste of horseradish (*wasabi*), and is served with soy sauce for dipping.
    Maki zushi contains strips of fish or vegetables rolled in rice and
wrapped in thin sheets of *nori*, or dried seaweed. Popular ingredients
include smoked salmon, fresh crab, shrimp, octopus, raw clams, and sea
urchin. Americans have invented many of their own *maki zushi* combi-
(50) nations, including the California roll, which contains imitation crabmeat
and avocado. They have also made innovations in the construction of
maki zushi. Some American *sushi* bars switch the placement of *nori* and
rice, while others don't use *nori*, and instead roll the *maki zushi* in fish
roe. These colorful, crunchy eggs add to the visual and taste appeal of
(55) the dish.

18. According to the passage, what other food also gained popularity
    in the 1970s?
    a. salads
    b. pepperoni pizza
    c. fried chicken
    d. fast-food burgers
    e. fried rice

19. What was Yohei Hanaya's contribution to *sushi*?
    a. He pressed the fish and rice together in a box.
    b. He introduced the population of Edo to the dish.
    c. He smoked the fish before putting it on vinegared rice.
    d. He used *wasabi* to flavor it.
    e. He used raw fish.

20. According to the passage, what does *shime* mean?
    a. salmon
    b. shrimp
    c. marinated
    d. roe
    e. seaweed

21. All of the following can be explicitly answered by reading the passage EXCEPT

    a. What is the definition of the word *sushi*?

    b. Did Japan's economic status have a bearing on *sushi's* popularity?

    c. Have Americans adapted *sushi* to make it more in keeping with their tastes?

    d. Why do some Americans prefer *maki zushi* over *nigiri zushi*?

    e. What happens to fish when it is layered together with rice and left for a period of months?

22. The passage describes Americans' *sushi* consumption as

    a. more than it was five years ago.

    b. important when watching baseball.

    c. taking place primarily in their homes.

    d. a trend due to supermarket marketing.

    e. beginning for many in college.

23. In line 3, *unpalatable* most nearly means

    a. not visually appealing.

    b. not good tasting.

    c. bad smelling.

    d. too expensive.

    e. rough to the touch.

24. What happens when fish is pickled (line 29)?

    a. It becomes crisp.

    b. It turns green.

    c. It dissolves into the rice.

    d. It is preserved.

    e. It gets dry.

25. What would be the best name for *maki zushi* that has the placement of the rice and *nori* switched?

    a. rice ball

    b. *maki maki*

    c. *zushi* deluxe

    d. inside-out

    e. *wasabi sashimi*

Questions 26–33 are based on the following passages.

Both of these passages were adapted from high school newspaper editorials concerning reality television.

PASSAGE 1

(1)   There comes a time in every boy's life when he becomes a man. On this fateful day, he will be swept up and put on an island to compete for one million dollars. Then, this man will realize that money can't buy happiness. He will find his soul mate, as we all do, on national TV,
(5)   picking a woman out of a line of twenty. By then it will be time for him to settle down, move to the suburbs, make friends with the neighbors, and then refurbish the neighbors' house.

Welcome to real life. That is, real life as the television networks see it.

(10)   Reality TV is flawed in many ways, but the most obvious is in its name. It purports to portray reality, but no "reality" show has succeeded in this endeavor. Instead, Reality TV is an extension of fiction, and there are no writers who need to be paid. Television executives love it because it is so much cheaper to produce than any other type
(15)   of programming, and it's popular. But the truth is that there is little or no reality in Reality TV.

Do you sing in the shower while dreaming of getting your own record deal? There are a couple of shows made just for you. Audition, and make the cut, so some British guy who has never sung a note can
(20)   rip you to pieces on live television. Or maybe you're lonely and fiscally challenged, and dream of walking down the aisle with a millionaire? Real marriage doesn't involve contestants who know each other for a couple of days. The people on these shows seem to be more interested in how they look on camera than in the character of the person they
(25)   might spend the rest of their life with. Let's hope that isn't reality.

There are also about a dozen decorating shows. In one case, two couples trade rooms and redecorate for each other. The catch is, interior designers help them. This is where the problem starts. Would either couple hire someone who thinks it's a great idea to swathe a
(30)   room in hundreds of yards of muslin, or to adhere five thousand plastic flowers as a mural in a bathroom? The crimes committed against defenseless walls are outrageous. When you add the fact that the couples are in front of cameras as well as the designers, and thus unable to react honestly to what is going on, you get a new level of "unreality."

(35)   Then there is the show that made the genre mainstream—*Survivor*. The show that pits men and women from all walks of life against each other for a million dollar prize in the most successful of all the Reality TV programs. What are record numbers of viewers tuning in to see? People who haven't showered or done their laundry in weeks are

(40)   shown scavenging for food and competing in ridiculous physical challenges. Where's the reality? From the looks of it, the contestants spend most of their time, when not on a Reality TV show, driving to the Burger Barn and getting exercise only when the remote goes missing.

So the television networks have used Reality TV to replace the dra-

(45)   mas and comedies that once filled their schedules, earning millions in advertising revenue. The lack of creativity, of producing something worth watching, is appalling. We are served up hundreds of hours of Reality TV each week, so we can watch real people in very unreal situations, acting as little like themselves as possible. What's real about that?

PASSAGE 2

(1)   Why does Reality TV get such a bad rap? Editorials on the subject blame its popularity on everything from the degenerate morals of today's youth to our ever-decreasing attention spans. The truth is that reality-based programs have been around for decades. *Candid Camera*

(5)   first aired in 1948, a "Cops"-like show called *Wanted* was on CBS's lineup in the mid-1950s, and PBS aired a controversial 12–hour documentary filmed inside a family's home in 1973. But it was *Survivor*, which debuted on American TV in the summer of 2000, which spawned the immense popularity of the "reality" genre. There are now

(10)   more than 40 reality shows on the air, and, hinting that they are here to stay, the Academy of Television Arts and Sciences added "Best Reality Show" as an Emmy category in 2002.

Why are these shows so popular today? Are they really a sign that our morals, and our minds, are on a decline? People have been tuning

(15)   in to Reality TV for generations, so what makes today's shows any worse than their predecessors? Let's look at a number of current, popular shows to see what the fuss is about. MTV's *The Real World* has been on the air for over ten years. It places seven strangers in one house and tapes them as they live together for a few months. The show has been

(20)   a ratings homerun for MTV, and tens of thousands of hopefuls audition each time they announce they are producing another show. Those who make the cut are attractive young singles not only looking for a good time, but also looking for fame, too. It's not uncommon for them to hire a show business agent before the taping starts.

(25)   Other Reality shows take fame-seekers to the next level by having them compete against one another. *American Idol, Star Search,* and *Fame* showcase singers, actors, dancers, and model wannabes, and offer them a chance at professional success. Even those who don't win the big prize get national television exposure, and have a better chance

(30)   than they did before the show of becoming famous. *Survivor* offers another twist: not only can you become an instant celebrity, but you have a chance to win a million dollars. The combination of fame and money has helped to make *Survivor* the most popular Reality TV program of all time. But it's not alone in the format. *Big Brother* combines

(35)   the "group living together in a beautiful setting" concept of *The Real World* with a $500,000 prize, and *Fear Factor* pays $50,000 to the contestant who completes the most terrifying stunts.

Given television's long history of reality-based programming, why is there a problem now? Most Reality TV centers on two common

(40)   motivators: fame and money. The shows have pulled waitresses, hair stylists, investment bankers, and counselors, to name a few, from obscurity to household names. These lucky few successfully parlayed their fifteen minutes of fame into celebrity. Even if you are not interested in fame, you can probably understand the desire for lots of

(45)   money. Watching people eat large insects, jump off cliffs, and be filmed 24 hours a day for a huge financial reward makes for interesting viewing. What's wrong with people wanting to be rich and famous? Not much, and, if you don't like it, you can always change the channel.

26. The author's tone in Passage 1, lines 1–7, may best be described as
   a. satire concerning a man's journey through life.
   b. cynicism about the reasons people go on Reality TV shows.
   c. humor regarding the content of Reality TV.
   d. irony about the maturation process.
   e. sarcasm toward the television networks.

27. Based on the passages, which statement would both authors agree with?
   a. Reality TV has had a long history.
   b. *Big Brother* is about the desire for fame and money.
   c. The popularity of Reality TV is an indication of a decline in morals.
   d. *Survivor* is the most successful Reality TV show.
   e. There is nothing wrong with Reality TV.

28. The primary purpose of Passage 2 is to
   a. refute an argument.
   b. explore possible outcomes.
   c. give a brief history.
   d. explain how to get famous.
   e. show the need for change.

29. The two passages differ in that the author of Passage 1
   a. defends Reality TV, while the author of Passage 2 does not.
   b. explains what he or she thinks is wrong with Reality TV, while the author of Passage 2 does not.
   c. believes Reality TV has many faults, while the author of Passage 2 thinks no one has a problem with it.
   d. blames Reality TV for the lack of variety in programming, while the author of Passage 2 thinks it has improved variety.
   e. says Reality TV is cheap to produce, while the author of Passage 2 disagrees.

30. In Passage 2, line 20, the phrase *ratings homerun* means that
   a. a lot of people watch *The Real World*.
   b. *The Real World* beats baseball games in TV ratings.
   c. there are baseball players on *The Real World*.
   d. the Nielsen company likes *The Real World*.
   e. *The Real World* contestants play softball on the show.

31. Both passages illustrate the idea that
   a. people on Reality TV shows become famous.
   b. Reality TV is all about getting rich.
   c. Reality TV is a good alternative to traditional programming.
   d. the producers of Reality TV are getting rich.
   e. Reality TV is controversial.

32. *Swathe* in Passage 1, line 29 most nearly means
   a. to stitch.
   b. a combination of pleating and stapling.
   c. to cover.
   d. a way of making curtains.
   e. to cover the floor.

33. What does the author of Passage 1 find most troublesome about Reality TV?
   a. It isn't original.
   b. It doesn't need writers to come up with scripts.
   c. It invades people's privacy.
   d. It doesn't accurately show reality.
   e. It shows how shallow people are.

Questions 34–40 are based on the following passage.

The selection that follows is based on an excerpt from a history of the game of Monopoly.

(1) In 1904, the U.S. Patent Office granted a patent for a board game called "The Landlord's Game," which was invented by a Virginia Quaker named Lizzie Magie. Magie was a follower of Henry George, who started a tax movement that supported the theory that the rent-

(5) ing of land and real estate produced an unearned increase in land values that profited a few individuals (landlords) rather than the majority of the people (tenants). George proposed a single federal tax based on land ownership; he believed this tax would weaken the ability to form monopolies, encourage equal opportunity, and narrow the gap

(10) between rich and poor.

Lizzie Magie wanted to spread the word about George's proposal, making it more understandable to a majority of people who were basically unfamiliar with economics. As a result, she invented a board game that would serve as a teaching device. The Landlord's Game was

(15) intended to explain the evils of monopolies, showing that they repressed the possibility for equal opportunity. Her instructions read in part: "The object of this game is not only to afford amusement to players, but to illustrate to them how, under the present or prevailing system of land tenure, the landlord has an advantage over other enter-

(20) prisers, and also how the single tax would discourage speculation."

The board for the game was painted with forty spaces around its perimeter, including four railroads, two utilities, twenty-two rental properties, and a jail. There were other squares directing players to go to jail, pay a luxury tax, and park. All properties were available for rent,

(25) rather than purchase. Magie's invention became very popular, spreading through word of mouth, and altering slightly as it did. Since it was not manufactured by Magie, the boards and game pieces were homemade. Rules were explained and transmuted, from one group of friends

(30) to another. There is evidence to suggest that The Landlord's Game was played at Princeton, Harvard, and the University of Pennsylvania.

In 1924, Magie approached George Parker (President of Parker Brothers) to see if he was interested in purchasing the rights to her game. Parker turned her down, saying that it was too political. The game increased in popularity, migrating north to New York state, west
(35) to Michigan, and as far south as Texas. By the early 1930s, it reached Charles Darrow in Philadelphia. In 1935, claiming to be the inventor, Darrow got a patent for the game, and approached Parker Brothers. This time, the company loved it, swallowed Darrow's prevarication, and not only purchased his patent, but paid him royalties for every
(40) game sold. The game quickly became Parker Brothers' bestseller, and made the company, and Darrow, millions of dollars.

When Parker Brothers found out that Darrow was not the true inventor of the game, they wanted to protect their rights to the successful game, so they went back to Lizzie Magie, now Mrs. Elizabeth
(45) Magie Phillips of Clarendon, Virginia. She agreed to a payment of $500 for her patent, with no royalties, so she could stay true to the original intent of her game's invention. She therefore required in return that Parker Brothers manufacture and market The Landlord's Game in addition to Monopoly. However, only a few hundred games
(50) were ever produced. Monopoly went on to become the world's bestselling board game, with an objective that is the exact opposite of the one Magie intended: "The idea of the game is to buy and rent or sell property so profitably that one becomes the wealthiest player and eventually monopolist. The game is one of shrewd and amusing trad-
(55) ing and excitement."

34. In line 16, what does *repressed the possibility for equal opportunity* mean?
   a. Monopolies led to slavery.
   b. Monopolies were responsible for the single tax problems.
   c. Monopolies made it impossible for poorer people to follow Henry George.
   d. Monopolies were responsible for Lizzie Magie's $500 payment and Charles Darrow's millions.
   e. Monopolies made it impossible for poorer people to have the same chances as the wealthy.

35. How does the objective of The Landlord's Game differ from that of Monopoly?

    a. In The Landlord's Game, you can only rent the properties, but in Monopoly you may buy them.

    b. The Landlord's Game illustrates the inequality of the landlord/tenant system, while Monopoly encourages players to become landlords and become wealthy at the expense of others.

    c. The Landlord's Game teaches the problems of capitalism and Monopoly teaches the value of money.

    d. The Landlord's Game was a way for Quakers to understand the economic theories of Henry George, and Monopoly explains the evolutionary theories of Charles Darrow.

    e. In The Landlord's Game, players try to land on as many railroads and utilities as possible, but in Monopoly they try to avoid them.

36. In line 38, what does *swallowed Darrow's prevarication* mean?

    a. ate his lunch

    b. believed his lie

    c. understood his problem

    d. played by his rules

    e. drank his champagne

37. In line 28, the statement that the rules of The Landlord's Game were *explained and transmuted* relies on the notion that

    a. when people pass along information by word of mouth, it goes through changes.

    b. when people explain things to their friends, they take on a different appearance.

    c. friends rely on one another for vital information.

    d. it's not always easy to play by the rules.

    e. word of mouth is the best way to spread information.

38. In paragraph 4, the author implies that

    a. Parker Brothers bought the game from Charles Darrow.

    b. it is not difficult to get a patent for an idea you didn't invent.

    c. Monopoly made Parker Brothers and Darrow millions of dollars.

    d. Lizzie Magie tried to sell her game to George Parker.

    e. The Landlord's Game was popular with Quakers.

39. Why did Mrs. Phillips sell her patent to Parker Brothers?
   a. So a large company would market her game and spread the word about Henry George's single tax theory.
   b. So she could make money.
   c. So The Landlord's Game could compete with Monopoly.
   d. So the truth would be told about Charles Darrow.
   e. So she would become famous.

40. All of the following questions can be explicitly answered on the basis of the passage EXCEPT
   a. Why did Lizzie Magie invent The Landlord's Game?
   b. Was was the object of The Landlord's Game?
   c. What were some of the properties on The Landlord's Game board?
   d. Who did Charles Darrow sell the game to?
   e. How did Parker Brothers find out that Charles Darrow didn't invent the game?

Questions 41–47 are based on the following passage.

The following selection is adapted from a news story about a bill recently introduced in Congress.

(1) In the past thirty years, Americans' consumption of restaurant and take-out food has doubled. The result, according to many health watchdog groups, is an increase in overweight and obesity. Almost 60 million Americans are obese, costing $117 billion each year in health

(5) care and related costs. Members of Congress have decided they need to do something about the obesity epidemic. A bill was recently introduced in the House that would require restaurants with twenty or more locations to list the nutritional content of their food on their menus. A Senate version of the bill is expected in the near future.

(10) Our legislators point to the trend of restaurants' marketing larger meals at attractive prices. People order these meals believing that they are getting a great value, but what they are also getting could be, in one meal, more than the daily recommended allowances of calories, fat, and sodium. The question is, would people stop "supersizing," or

(15) make other healthier choices if they knew the nutritional content of the food they're ordering? Lawmakers think they would, and the gravity of the obesity problem has caused them to act to change menus.

The Menu Education and Labeling, or MEAL, Act, would result in menus that look like the nutrition facts panels found on food in super-

(20) markets. Those panels are required by the 1990 Nutrition Labeling

and Education Act, which exempted restaurants. The new restaurant menus would list calories, fat, and sodium on printed menus, and calories on menu boards, for all items that are offered on a regular basis (daily specials don't apply). But isn't this simply asking restaurants to
(25) state the obvious? Who isn't aware that an order of supersize fries isn't health food? Does anyone order a double cheeseburger thinking they're being virtuous?

Studies have shown that it's not that simple. In one, registered dieticians couldn't come up with accurate estimates of the calories found in
(30) certain fast foods. Who would have guessed that a milk shake, which sounds pretty healthy (it does contain milk, after all) has more calories than three McDonald's cheeseburgers? Or that one chain's chicken breast sandwich, another better-sounding alternative to a burger, contains more than half a day's calories and twice the recommended daily
(35) amount of sodium? Even a fast-food coffee drink, without a doughnut to go with it, has almost half the calories needed in a day.

The restaurant industry isn't happy about the new bill. Arguments against it include the fact that diet alone is not the reason for America's obesity epidemic. A lack of adequate exercise is also to blame. In addi-
(40) tion, many fast food chains already post nutritional information on their websites, or on posters located in their restaurants.

Those who favor the MEAL Act, and similar legislation, say in response that we must do all we can to help people maintain a healthy weight. While the importance of exercise is undeniable, the quantity
(45) and quality of what we eat must be changed. They believe that if we want consumers to make better choices when they eat out, nutritional information must be provided where they are selecting their food. Restaurant patrons are not likely to have memorized the calorie counts they may have looked up on the Internet, nor are they going to leave
(50) their tables, or a line, to check out a poster that might be on the opposite side of the restaurant.

41. The purpose of the passage is to
    a. targue the restaurant industry's side of the debate.
    b. explain why dieticians have trouble estimating the nutritional content of fast food.
    c. help consumers make better choices when dining out.
    d. explain one way legislators propose to deal with the obesity epidemic.
    e. argue for the right of consumers to understand what they are ordering in fast food restaurants.

42. According to the passage, the larger meals now being offered in restaurants
  a. cost less than smaller meals.
  b. add an extra side dish not offered with smaller meals.
  c. include a larger drink.
  d. save consumers money.
  e. contain too many calories, fat, and sodium.

43. In lines 15–16, the word *gravity* most nearly means
  a. the force of attraction toward earth.
  b. a cemetery plot.
  c. seriousness.
  d. jealousy.
  e. presumption of wrongdoing.

44. According to the passage, why is the restaurant industry against the new Congressional bill?
  a. They don't want any healthy items on their menus.
  b. Because lack of adequate exercise is also responsible for the obesity epidemic.
  c. They don't want to be sued if they incorrectly calculate the calories in their menu items.
  d. They feel their industry is already over-regulated.
  e. Because people would stop coming to their establishments if they knew what was in the food.

45. Why is the chicken breast sandwich mentioned in paragraph 4?
  a. It is an example of a menu item that contains more fat than one would assume.
  b. It is the only healthy choice on some restaurants' menus.
  c. It has twice as much salt as the recommended daily allowance.
  d. It has as many calories as three McDonald's hamburgers.
  e. It is a typical selection in a Value Meal.

46. The passage explains that those in favor of the MEAL Act want nutritional information placed
  a. anywhere the consumer can make a menu selection.
  b. in print advertisements.
  c. on websites.
  d. on toll-free hotlines.
  e. on posters with print large enough to read from any position in the restaurant.

47. If the MEAL Act is passed, consumers would see
   a. menus that tell them how to select the healthiest complete meal.
   b. menus that look like nutritional labels on packaged food.
   c. restaurants with more extensive information on their websites.
   d. less television advertising of fast food restaurants.
   e. restaurants that serve healthier food choices.

Answers

1. **c.** The answer may be found in lines 4 and 5, which state that Russell wanted *an alternative to his scratched and warped phonograph records*. You may infer that the problem with such records was their poor sound quality.

2. **e.** Lines 26–27 state that the detector's function is to convert data collected by the laser into music.

3. **b.** While the paragraph explains the function of semiconductor lasers in reading the information on CDs, it does not say anything about why they were invented.

4. **a.** Evidence may be found in lines 23–24, which state that today's models are quirkier and less perfect than the supermodels.

5. **c.** A resumé is literally the summary of one's job experience, education, and skills. The author is saying that there is nothing one can say about these models except that they look great; their figurative resumé has only one item on it. Being great-looking isn't work experience (choice **a**), one would not literally list "great-looking" alone on a resume (choices **b** and **d**), and *pathos* is a feeling of pity or sorrow (choice **e**).

6. **e.** To *wax* means to become, and *rhapsodic* means excessively enthusiastic. Although rhapsodic can also mean like a musical composition of irregular form, this definition does not fit with the rest of the sentence.

7. **e.** Lines 5–7 mention calculators (adding machines), computers, card punches, and manuals. The only item not mentioned is kitchen scales.

8. **c.** A sneer is a facial expression that signals contempt or scorn. Accountants and bookkeepers didn't like the comptometer, because as lines 13–14 explain, it performed their job faster than they could.

9. **b.** The Museum has a collection of computer-related magazines, manuals, and books (line 7). They would not contain information on the inventor of the telephone (choice **a**), other museums in California (choice **c**), the profession of comptometer operation (choice **d**), or why video games are harmful (choice **e**). Since IBM played, and continues to play, an important role in the development of computers and computer-related technology, it could most likely be researched at the Museum.

10. d. Lines 4–5 explain that there was a social component to a trip to the marketplace. To be social means to be around others, suggesting that people sought out interaction with one another.

11. c. The prefix *ante-* means earlier, as does *pre-*. Additional context clues may be found in the first paragraph, which explains the similarities between historical marketplaces (those of long ago), and the malls of today, and in line 6, which states the mall is a descendant of the marketplace.

12. a. This information is not given in the passage.

13. b. The answer is in lines 27–29: *It was constructed according to a unified plan, rather than as a random group of stores. Nichols' company owned and operated the mall, leasing space to a variety of tenants.*

14. e. Lines 31–34 explain that Gruen took the shopping mall to the next level by intending it to take the place of a city center, with leisure and entertainment opportunities as well as shopping and dining.

15. b. All of the other choices are mentioned in lines 46–48.

16. a. Lines 36–38 list some of Southdale's offerings, such as shops, restaurants, a school, a post office, a skating rink, works of art, and fountains. These are also available in a city, and may be considered among the pleasures of urban life.

17. e. All of the other choices were mentioned in the last two paragraphs as positive impacts of megamalls. However it is unlikely that a mall in Minnesota would be in direct competition for visitors with a Mall located on the other side of the world.

18. a. Salad is the best choice, because (lines 4–7) at the time, Americans were beginning to eat healthier foods, such as vegetables.

19. e. Lines 33 and 34 explain that he skipped the fermentation process, which means that the fish was fresh, or raw. If you answered choice **b**, check back to the passage. There is no reason to believe that sushi with fermented rice was not being consumed in Edo before Yohei's innovation. If you answered choice **d**, note that the passage does not indicate when, or with whom, *wasabi* began being used as a condiment with *nigiri zushi*.

20. c. It states in lines 42 and 43 that *ama ebi* is raw shrimp, and *shime saba* is marinated mackerel. You can infer that *ebi* means shrimp, because "raw" is not one of your choices. You can also infer that *shime* means marinated, because mackerel is not one of your choices. Therefore, *shime ebi* means marinated shrimp.

21. d. Nowhere in the passage does the author mention a preference for either type of sushi. The answer to choice **a** may be found in lines 36 and 37. Choice **b** is found in lines 10–13, choice **c** is answered by lines 46–51, and choice **e** is answered by lines 26–29.

22. a. It is noted in lines 15 and 16 that sushi consumption in America is 40% higher than it was in the late 1990s (five years ago). While the other answers might be true, they are not described in the passage.

23. b. *Unpalatable* may be defined as not agreeable to taste; from the Latin *palatum*, which refers to the roof of the mouth. You know the word *palate* as the roof of the mouth, so unpalatable most likely has to do with the sense of taste. The biggest clue to the definition comes in line 24, which states that Americans have decided, *this once-scorned food is truly delicious.*

24. d. It is mentioned in lines 25–26 that sushi was developed for the purpose of preserving fish. Line 29 clearly states that pickling, which takes place at the end of the sushi-making process, is a means of preserving.

25. d. The *nori* is typically on the outside of the roll, surrounding the rice (lines 46 and 47). If the rice is wrapped around the seaweed, the inside (rice) is now on the outside. In addition, you could use the process of elimination, as none of the other choices make sense.

26. c. The author does not have a bite to his argument, as required by satire, cynicism, and sarcasm. He is also not speaking to two audiences, one that *gets it* and one that doesn't, as with irony. He is simply trying to be funny, as in lines 1–3, which says that once a boy becomes a man, he will compete for cash on an island.

27. d. This is the only statement made by both authors (see Passage 1 lines 37–38, and Passage 2 lines 33–34). Don't be tricked by the choices that are true, such as **a**, **b**, and **e**. They need to be believed by both authors to be correct.

28. a. Passage 2 repeats a number of times its first question: Why does Reality TV get such a bad rap? Lines 2 and 3 explain the argument further, saying its popularity is blamed on degenerate morals and a decreasing attention span. The first lines of paragraph 2 (13–16) again question the argument against Reality

TV, and the last paragraph repeats the questioning. There are no outcomes or any need for change mentioned. A brief history is given, and the subject of getting famous through exposure on Reality TV is brought up, but neither is the primary purpose of the passage.

29. b. Passage 1 centers on a problem with Reality TV, and while Passage 2 does mention some problems, they are not what he or she feels, but rather the opinion of *some people*. Choice **a** is incorrect because Passage 1 does not defend Reality TV. Choice **c** is incorrect because the author of Passage 2 acknowledges that some people have a problem with Reality TV (lines 1–3 and 48–49). Choice **d** is incorrect because Passage 2 does not say anything about variety in TV programming. Choice **e** is wrong because Passage 2 doesn't mention the cost of producing TV shows.

30. a. Ratings refers to how many people watch the show. A homerun is the best possible kind of hit, so a *ratings homerun* is a symbolic term meaning that many people watch the show. Choices **b, c,** and **e** reference ball games literally, but the author used the term figuratively, so those choices are incorrect. Nielsen is the company that gathers TV ratings, but high ratings have nothing to do with whether they like a show or not.

31. e. Both passages show that there is a debate about Reality TV. In Passage 1, the author is against it, but notes that it is popular (lines 10 and 37). The author of Passage 2 likes it, and also recognizes that it gets a *bad rap* (line 1). Although most of the other choices are factual, they do not appear in both passages, and are not illustrated by them.

32. c. The clue comes in Passage 1, which describes the swathing and flower gluing as crimes against defenseless walls. Swathing is therefore something done to a wall. The only choice that makes sense is **c**, to cover.

33. d. While there is evidence for the other choices, they are not the most troublesome. The author repeats in every paragraph the idea that Reality TV isn't real.

34. e. Look back to lines 7–10, where George's single tax proposal (the idea The Landlord's Game was meant to teach) is described as aiming to *weaken the ability to form monopolies, encourage equal opportunity, and narrow the gap between rich and poor.*

35. b. Lines 13–20 explain the first part of the question, while lines 52–55 contain the answer to the second. Don't be distracted by the other answers that contain true statements that are not, however, the objectives of the games. Note also that evolution was a theory of Charles Darwin, not Charles Darrow.

36. b. Lines 35–37 explains that Darrow fraudulently claimed to be the game's inventor (he was introduced to it before he got a patent as its inventor). Parker Brothers bought his patent believing that it was genuine, meaning that they believed Darrow's falsehood.

37. a. The answer is in line 26. Having the game and its rules spread by word of mouth means it will *alter slightly* from one person to another.

38. b. To *imply* means to hint at, rather than to state outright. The other choices are all directly stated in the paragraph, while **b** is implied.

39. a. Lines 46 and 47 say she sold it to remain true to her original intent, which was, according to line 11, to spread the word about George's single tax theory.

40. e. Lines 42 and 43 say that Parker Brothers found out that Darrow wasn't the inventor, but nowhere in the passage does it say how they learned the information.

41. d. In the first paragraph, where the theme is typically introduced, it states that *members of Congress have decided they need to do something about the obesity epidemic* (lines 5 and 6).

42. e. The answer is found in lines 12–14: *what they are also getting could be, in one meal, more than the daily recommended allowances of calories, fat, and sodium.*

43. c. Clues for this question are found in the first paragraph, in which the obesity problem is called an epidemic, and the staggering cost of the problem is mentioned.

44. b. Paragraph 5 states that the restaurant industry has responded to the bill by pointing out that *diet alone is not the reason for America's obesity epidemic. A lack of adequate exercise is also to blame.*

45. c. The answer is in lines 32–35: the chicken breast sandwich contains more than *twice the recommended daily amount of sodium.*

46. a. Paragraph 6 explains that those who support the MEAL Act believe *nutritional information must be provided where they are selecting their food* (lines 46 and 47).

47. b. The answer is in lines 18–20: *The Menu Education and Labeling, or MEAL, Act, would result in menus that look like the nutrition facts panels found on food in supermarkets.*

U.S. History and Politics

Questions 48–51 are based on the following passage.

The following passage discusses the Supreme Court's power of judicial review, a practice first invoked in the historical 1803 Supreme Court case Marbury v. Madison.

(1)   "It is emphatically the province and duty of the judicial department to say what the law is," stated Chief Justice John Marshall in a unanimous opinion in the 1803 Supreme Court case of *Marbury v. Madison*. This landmark case established the doctrine of judicial review, which gives

(5)   the court the authority to declare executive actions and laws invalid if they conflict with the U.S. Constitution. The court's ruling on the constitutionality of a law is nearly final—it can only be overcome by a constitutional amendment or by a new ruling of the court. Through the power of judicial review, the court shapes the development of law,

(10)   assures individual rights, and maintains the Constitution as a "living" document by applying its broad provisions to complex new situations.

Despite the court's role in interpreting the Constitution, the document itself does not grant this authority to the court. However, it is clear that several of the founding fathers expected the Court to act in

(15)   this way. Alexander Hamilton and James Madison argued for the importance of judicial review in the *Federalist Papers*, a series of 85 political essays that urged the adoption of the Constitution. Hamilton

argued that judicial review protected the will of the people by making the Constitution supreme over the legislature, which might only (20) reflect the temporary will of the people. Madison wrote that if a public political process determined the constitutionality of laws, the Constitution would become fodder for political interests and partisanship. However, the practice of judicial review was, and continues to be, a controversial power because it gives justices—who are appointed (25) rather than elected by the people—the authority to void legislation made by Congress and state lawmakers.

48. The passage suggests that the practice of judicial review allows the court to
   a. wield enormous power.
   b. determine foreign policy.
   c. make laws that reflect the principles of the Constitution.
   d. rewrite laws that are unconstitutional.
   e. make amendments to the Constitution.

49. The image of the Constitution as *a "living" document* (lines 10 and 11) implies that
   a. the supreme law of the land cannot be altered in any way.
   b. it can only be amended through a difficult process.
   c. its principles need to be adapted to contemporary life.
   d. the original document is fragile and needs to be preserved in the Library of Congress so that it will not deteriorate.
   e. it will die if it is interpreted by the court.

50. In line 5, *declare* most nearly means
   a. narrate.
   b. recite.
   c. proclaim.
   d. predict.
   e. acknowledge.

51. The last sentence (lines 23–26) in the passage provides
   a. a specific example supporting the argument made earlier.
   b. a summary of the points made earlier.
   c. an explanation of the positions made earlier.
   d. a prediction based on the argument made earlier.
   e. a counter-argument to the views referred to earlier.

Questions 52–55 are based on the following passage.

In the following passage, the author gives an account of the development of the Emancipation Proclamation, Abraham Lincoln's 1863 executive order abolishing slavery in the Confederate States of America.

(1)   Almost from the beginning of his administration, Lincoln was pressured by abolitionists and radical Republicans to issue an Emancipation Proclamation. In principle, Lincoln approved, but he postponed action against slavery until he believed he had wider support from the
(5)   American public. The passage of the Second Confiscation Act by Congress on July 17, 1862, which freed the slaves of everyone in rebellion against the government, provided the desired signal. Not only had Congress relieved the Administration of considerable strain with its limited initiative on emancipation, it demonstrated an increasing pub-
(10)   lic abhorrence toward slavery. Lincoln had already drafted what he termed his "Preliminary Proclamation." He read his initial draft of the Emancipation Proclamation to Secretaries William H. Seward and Gideon Welles on July 13, 1862. For a moment, both secretaries were speechless. Quickly collecting his thoughts, Seward said something
(15)   about anarchy in the South and possible foreign intervention, but with Welles apparently too confused to respond, Lincoln let the matter drop.

   Nine days later, on July 22, Lincoln raised the issue in a regularly scheduled Cabinet meeting. The reaction was mixed. Secretary of War
(20)   Edwin M. Stanton, correctly interpreting the Proclamation as a military measure designed both to deprive the Confederacy of slave labor and bring additional men into the Union Army, advocated its immediate release. Treasury Secretary Salmon P. Chase was equally supportive, but Montgomery Blair, the Postmaster General, foresaw
(25)   defeat in the fall elections. Attorney General Edward Bates, a conservative, opposed civil and political equality for blacks but gave his qualified support. Fortunately, President Lincoln only wanted the advice of his Cabinet on the style of the Proclamation, not its substance. The course was set. The Cabinet meeting of September 22, 1862, resulted
(30)   in the political and literary refinement of the July draft, and on January 1, 1863, Lincoln composed the final Emancipation Proclamation. It was the crowning achievement of his administration.

52. The passage suggests which of the following about Lincoln's Emancipation Proclamation?
  a. Abolitionists did not support such an executive order.
  b. The draft proclamation was unanimously well-received by Lincoln's cabinet.
  c. Congressional actions influenced Lincoln and encouraged him to issue it.
  d. The proclamation was not part of a military strategy.
  e. The first draft needed to be edited because Lincoln made numerous grammatical errors.

53. The description of the reaction of Secretaries Seward and Welles to Lincoln's draft proclamation in lines 13–16 is used to illustrate
  a. Lincoln's lack of political acumen.
  b. that Lincoln's advisors did not anticipate his plan.
  c. the incompetence of Lincoln's advisors.
  d. Seward and Welles' disappointment that Lincoln did not free all slaves at that time.
  e. that most members of Lincoln's administration were abolitionists.

54. In lines 26 and 27, *qualified* most nearly means
  a. adept.
  b. capable.
  c. certified.
  d. eligible.
  e. limited.

55. The author's attitude to the issuing of the Emancipation Proclamation is one of
  a. informed appreciation.
  b. reluctant admiration.
  c. ambiguous acceptance.
  d. conflicted disapproval.
  e. personal dislike.

Questions 56–59 are based on the following passage.

The following passage describes the medium of political cartoons as a graphic means of commenting on contemporary social or political issues.

(1)   A mainstay of American newspapers since the early nineteenth century, political cartoons use graphic art to comment on current events in a

way that will inform, amuse, provoke, poke, and persuade readers. Cartoons take on the principal issues and leaders of the day, skewering hyp-
(5) ocritical or corrupt politicians and depicting the ridiculous, the ironic, or the serious nature of a major event in a single, deftly drawn image. Cartoons use few words, if any, to convey their message. Some use caricature, a technique in which a cartoonist exaggerates the features of well-known people to make fun of them. (Think of renderings of Bill
(10) Clinton with a nose redder than Rudolph's and swollen out of proportion, or cartoons of George W. Bush's exaggerated pointy visage sporting a ten-gallon cowboy hat.)

Because they have the ability to evoke an emotional response in readers, political cartoons can serve as a vehicle for swaying public
(15) opinion and can contribute to reform. Thomas Nast (1840–1902), the preeminent political cartoonist of the second half of the nineteenth century, demonstrated the power of his medium when he used his art to end the corrupt Boss Tweed Ring in New York City. His images, first drawn for *Harper's Weekly*, are still in currency today: Nast created
(20) the tiger as the symbol of Tammany Hall, the elephant for the Republican Party, and the donkey for the Democratic Party. Created under tight deadlines for ephemeral, commercial formats like newspapers and magazines, cartoons still manage to have lasting influence. Although they tackle the principal issues and leaders of their day, they
(25) often provide a vivid historical picture for generations to come.

56. The author would most likely agree with which statement?
   a. Political cartoons are a powerful means of influencing the public.
   b. The more mean-spirited a political cartoon is, the more effective.
   c. Political cartoonists must maintain their objectivity on controversial subjects.
   d. Political cartoons cater to an elite class of intellectuals.
   e. Because of their relevance to current affairs, political cartoons rarely serve as historical documents.

57. In describing the art of political cartooning in the first paragraph, the author's tone can be best described as
   a. sober.
   b. earnest.
   c. critical.
   d. impartial.
   e. playful.

58. In line 14, *vehicle* most nearly means
  a. automobile.
  b. carrier.
  c. tunnel.
  d. outlet.
  e. means.

e

59. The author cites Thomas Nast's depiction of an elephant for the Republican Party (lines 20–21) as an example of
  a. an image that is no longer recognized by the public.
  b. the saying "the pen is mightier than the sword."
  c. art contributing to political reform.
  d. a graphic image that became an enduring symbol.
  e. the ephemeral naature of political cartooning.

d

2/9/19   12-30-13 9/26/18

Questions 60–67 are based on the following passage.

Beginning in the 1880s, southern states and municipalities established statutes called Jim Crow laws that legalized segregation between blacks and whites. The following passage is concerned with the fight against racial discrimination and segregation and the struggle for justice for African Americans in post-World War II United States.

(1)   The post-World War II era marked a period of unprecedented energy against the second-class citizenship accorded to African Americans in many parts of the nation. Resistance to racial segregation and discrimination with strategies like those described above—civil disobe-

(5)   dience, nonviolent resistance, marches, protests, boycotts, "freedom rides," and rallies—received national attention as newspaper, radio, and television reporters and cameramen documented the struggle to end racial inequality.

When Rosa Parks refused to give up her seat to a white person in

(10) Montgomery, Alabama, and was arrested in December 1955, she set off a train of events that generated a momentum the civil rights movement had never before experienced. Local civil rights leaders were hoping for such an opportunity to test the city's segregation laws. Deciding to boycott the buses, the African-American community soon

(15) formed a new organization to supervise the boycott, the Montgomery Improvement Association (MIA). The young pastor of the Dexter Avenue Baptist Church, Reverend Martin Luther King, Jr., was chosen as the first MIA leader. The boycott, more successful than anyone

(20) hoped, led to a 1956 Supreme Court decision banning segregated buses.

In 1960, four black freshmen from North Carolina Agricultural and Technical College in Greensboro strolled into the F. W. Woolworth store and quietly sat down at the lunch counter. They were not served, but they stayed until closing time. The next morning they came with
(25) twenty-five more students. Two weeks later similar demonstrations had spread to several cities, within a year similar peaceful demonstrations took place in over a hundred cities North and South. At Shaw University in Raleigh, North Carolina, the students formed their own organization, the Student Non-Violent Coordinating Committee
(30) (SNCC, pronounced "Snick"). The students' bravery in the face of verbal and physical abuse led to integration in many stores even before the passage of the Civil Rights Act of 1964.

The August 28, 1963, March on Washington riveted the nation's attention. Rather than the anticipated hundred thousand marchers,
(35) more than twice that number appeared, astonishing even its organizers. Blacks and whites, side by side, called on President John F. Kennedy and the Congress to provide equal access to public facilities, quality education, adequate employment, and decent housing for African Americans. During the assembly at the Lincoln Memorial, the
(40) young preacher who had led the successful Montgomery, Alabama, bus boycott, Reverend Dr. Martin Luther King, Jr., delivered a stirring message with the refrain, "I Have a Dream."

There were also continuing efforts to legally challenge segregation through the courts. Success crowned these efforts: the Brown decision
(45) in 1954, the Civil Rights Act of 1964, and the Voting Rights Act in 1965 helped bring about the demise of the entangling web of legislation that bound blacks to second class citizenship. One hundred years after the Civil War, blacks and their white allies still pursued the battle for equal rights in every area of American life. While there is more
(50) to achieve in ending discrimination, major milestones in civil rights laws are on the books for the purpose of regulating equal access to public accommodations, equal justice before the law, and equal employment, education, and housing opportunities. African Americans have had unprecedented openings in many fields of learning and
(55) in the arts. The black struggle for civil rights also inspired other liberation and rights movements, including those of Native Americans, Latinos, and women, and African Americans have lent their support to liberation struggles in Africa.

60. The passage is primarily concerned with
  a. enumerating the injustices that African Americans faced.
  b. describing the strategies used in the struggle for civil rights.
  c. showing how effective sit-down strikes can be in creating change.
  d. describing the nature of discrimination and second class citizenship.
  e. recounting the legal successes of the civil rights movement.

61. The author cites the example of Rosa Parks (lines 9–10) refusing to relinquish her bus seat in order to
  a. demonstrate the accidental nature of political change.
  b. show a conventional response to a common situation.
  c. describe a seminal event that influenced a larger movement.
  d. portray an outcome instead of a cause.
  e. give a detailed account of what life was like in Montgomery, Alabama in 1955.

62. In line 13, the word *test* most nearly means
  a. analyze.
  b. determine.
  c. prove.
  d. quiz.
  e. challenge.

63. The passage suggests that the college students in Greensboro, North Carolina (lines 21–27)
  a. were regulars at the Woolworth lunch counter.
  b. wanted to provoke a violent reaction.
  c. were part of an ongoing national movement of lunch-counter demonstrations.
  d. inspired other students to protest peacefully against segregation.
  e. did not plan to create a stir.

64. The passage implies that the 1963 March on Washington
  a. resulted in immediate legislation prohibiting segregation in public accommodations.
  b. was a successful demonstration that drew attention to its causes.
  c. was overshadowed by the rousing speech by Dr. Martin Luther King, Jr.
  d. represented only the attitudes of a fringe group.
  e. reflected unanimous public opinion that segregation laws must end.

65. The term *refrain* as it is used in line 42 most nearly means
    a. song lyric.
    b. allegory.
    c. recurring phrase.
    d. poem stanza.
    e. aria.

66. The term *second class citizenship* (line 47) most nearly refers to
    a. native or naturalized people who do not owe allegiance to a government.
    b. foreign-born people who wish to become a citizen of a new country.
    c. those who deny the rights and privileges of a free person.
    d. having inferior status and rights in comparison to other citizens.
    e. having inferior status and rights under a personal sovereign.

67. All of the following questions can be explicitly answered on the basis of the passage EXCEPT
    a. What are some of the barriers African Americans faced in post-war America?
    b. What tangible achievements did the civil rights movement attain?
    b. What judicial rulings are considered milestones in the struggle for civil rights?
    b. What strategies did civil rights protesters use to provoke political change?
    b. What hurtles remain today for ending racial discrimination in the United States?

Questions 68–75 are based on the following passage.

The following passage explores the role of Chinese Americans in the nineteenth-century westward expansion of the United States, specifically their influence on the development of California.

(1)   While the Chinese, in particular those working as sailors, knew the west coast of North America before the Gold Rush, our story begins in 1850, as the documentation from the Gold Rush provides the starting point with which to build a more substantial narrative. Most Chinese immi-

(5)   grants entered California through the port of San Francisco. From San Francisco and other ports, many sought their fortunes in other parts of California. The Chinese formed part of the diverse gathering of peoples

from throughout the world who contributed to the economic and population explosion that characterized the early history of the state of Cal-
(10)   ifornia. The Chinese who emigrated to the United States at this time
were part of a larger exodus from southeast China searching for better
economic opportunities and fleeing a situation of political corruption and
decline. Most immigrants came from the Pearl River Delta in Guang-
dong (Canton) Province.

(15)   Chinese immigrants proved to be productive and resourceful con-
tributors to a multitude of industries and businesses. The initial group
of Chinese argonauts sought their livelihood in the gold mines, calling California *Gam Saan*, Gold Mountain. For the mining industry,
they built many of the flumes and roads, allowing for easier access and
(20)   processing of the minerals being extracted. Chinese immigrants faced
discrimination immediately upon arrival in California. In mining, they
were forced to work older claims, or to work for others. In the 1850s,
the United States Constitution reserved the right of naturalization for
white immigrants to this country. Thus, Chinese immigrants lived at
(25)   the whim of local governments with some allowed to become natu-
ralized citizens, but most not. Without this right, it was difficult to
pursue livelihoods. For example, Chinese immigrants were unable to
own land or file mining claims. Also in the 1850s, the California legislature passed a law taxing all foreign miners. Although stated in gen-
(30)   eral terms, it was enforced chiefly against the Mexicans and the
Chinese through 1870. This discrimination occurred in spite of the
fact that the Chinese often contributed the crucial labor necessary to
the mining enterprise.

Discriminatory legislation forced many Chinese out of the gold
(35)   fields and into low-paying, menial, and often arduous jobs. In many
cases, they took on the most dangerous and least desirable components of work available. They worked on reclaiming marshes in the
Central Valley so that the land could become agriculturally productive. They built the stone bridges and fences, constructed roads, and
(40)   excavated storage areas for the wine industry in Napa and Sonoma
counties. The most impressive construction feat of Chinese Americans
was their work on the western section of the transcontinental railroad.
Chinese-American workers laid much of the tracks for the Central
Pacific Railroad through the foothills and over the high Sierra
(45)   Nevada, much of which involved hazardous work with explosives to
tunnel through the hills. Their speed, dexterity, and outright perseverance, often in brutally cold temperatures and heavy snow through
two record breaking winters, is a testimony to their outstanding
achievements and contributions to opening up the West.

68. The first paragraph (lines 1–14) of the passage serves what function in the development of the passage?
   a. provides an expert's opinion to support the author's thesis
   b. introduces the topic by describing general patterns
   c. compares common myths with historical facts
   d. draws a conclusion about the impact of Chinese immigration on the state of California
   e. condemns outdated concepts

69. Which of the following best describes the approach of the passage?
   a. theoretical analysis
   b. historical overview
   c. dramatic narrative
   d. personal assessment
   e. description through metaphor

70. Lines 15–20 portray Chinese immigrants as
   a. fortuitous.
   b. prideful.
   c. vigorous.
   d. effusive.
   e. revolutionary.

71. The author cites the United States Constitution (lines 23–24) in order to
   a. praise the liberties afforded by the Bill of Rights.
   b. show that the government valued the contributions of its immigrants.
   c. imply that all American citizens are equal under the law.
   d. emphasize the importance of a system of checks and balances.
   e. suggest that it did not protect Chinese immigrants from discrimination.

72. The word *enterprise* as it is used in line 33 most nearly means
   a. organization.
   b. corporation.
   c. industry.
   d. partnership.
   e. occupation.

73. According to the passage, which of the following is NOT a contribution made by Chinese immigrants?
  a. worked land so that it would yield more crops
  b. performed dangerous work with explosives
  c. built roads and bridges
  d. purchased older mining claims and mined them
  e. dug storage areas for California wine

74. In line 37 *reclaiming* most nearly means
  a. redeeming.
  b. protesting.
  c. objecting.
  d. approving.
  e. extolling.

75. The last sentence (lines 46–49) in the passage provides
  a. an example supporting the thesis of the passage.
  b. a comparison with other historical viewpoints.
  c. a theory explaining historical events.
  d. a summary of the passage.
  e. an argument refuting the position taken earlier in the passage.

Questions 76–83 are based the following passage.

The following passage from the Lowell National Park Handbook describes the advent of American manufacturing, imported from England in the 1790s. The Arkwright system mentioned in the passage refers to a water frame, a water-powered spinning machine that was used to make cloth.

(1) The mounting conflict between the colonies and England in the 1760s and 1770s reinforced a growing conviction that Americans should be less dependent on their mother country for manufactures. Spinning bees and bounties encouraged the manufacture of homespun cloth as
(5) a substitute for English imports. But manufacturing of cloth outside the household was associated with relief of the poor. In Boston and Philadelphia, Houses of Industry employed poor families at spinning for their daily bread.

Such practices made many pre-Revolutionary Americans dubious
(10) about manufacturing. After independence there were a number of unsuccessful attempts to establish textile factories. Americans needed access to the British industrial innovations, but England had passed laws forbidding the export of machinery or the emigration of those who

(15)   could operate it. Nevertheless it was an English immigrant, Samuel
Slater, who finally introduced British cotton technology to America.

         Slater had worked his way up from apprentice to overseer in an
English factory using the Arkwright system. Drawn by American
bounties for the introduction of textile technology, he passed as a
farmer and sailed for America with details of the Arkwright water
(20)   frame committed to memory. In December 1790, working for mill
owner Moses Brown, he started up the first permanent American cotton spinning mill in Pawtucket, Rhode Island. Employing a workforce
of nine children between the ages of seven and twelve, Slater successfully mechanized the carding and spinning processes.

(25)   A generation of millwrights and textile workers trained under Slater
was the catalyst for the rapid proliferation of textile mills in the early
nineteenth century. From Slater's first mill, the industry spread across
New England to places like North Uxbridge, Massachusetts. For two
decades, before Lowell mills and those modeled after them offered
(30)   competition, the "Rhode Island System" of small, rural spinning mills
set the tone for early industrialization.

         By 1800 the mill employed more than 100 workers. A decade later
61 cotton mills turning more than 31,000 spindles were operating in
the United States, with Rhode Island and the Philadelphia region the
(35)   main manufacturing centers. The textile industry was established,
although factory operations were limited to carding and spinning. It
remained for Francis Cabot Lowell to introduce a workable power
loom and the integrated factory, in which all textile production steps
take place under one roof.

(40)   As textile mills proliferated after the turn of the century, a national
debate arose over the place of manufacturing in American society.
Thomas Jefferson spoke for those supporting the "yeoman ideal" of a
rural Republic, at whose heart was the independent, democratic
farmer. He questioned the spread of factories, worrying about factory
(45)   workers' loss of economic independence. Alexander Hamilton led
those who promoted manufacturing and saw prosperity growing out
of industrial development. The debate, largely philosophical in the
1790s, grew more urgent after 1830 as textile factories multiplied and
increasing numbers of Americans worked in them.

76. The primary purpose of the passage is to
   a. account for the decline of rural America.
   b. contrast political views held by the British and the Americans.
   c. summarize British laws forbidding the export of industrial machinery.
   d. describe the introduction of textile mills in New England.
   e. make an argument in support of industrial development.

77. The passage refers to Houses of Industry (line 7) to illustrate
   a. a highly successful and early social welfare program.
   b. the perception of cloth production outside the home as a social welfare measure.
   c. the preference for the work of individual artisans over that of spinning machines.
   d. the first textile factory in the United States.
   e. the utilization of technological advances being made in England at the time.

78. The first paragraph (lines 1–8) of the passage implies that early American manufacturing was
   a. entirely beneficial.
   b. politically and economically necessary.
   c. symbolically undemocratic.
   d. environmentally destructive.
   e. spiritually corrosive.

79. The description of Slater's immigration to the American colonies (lines 17–20) serves primarily to
   a. demonstrate Slater's craftiness in evading British export laws.
   b. show the attraction of farming opportunities in the American colonies.
   c. explain the details of British manufacturing technologies.
   d. illustrate American efforts to block immigration to the colonies.
   e. describe the willingness of English factories to share knowledge with the colonies.

80. Lines 22–24 imply that Slater viewed child labor as
   a. an available workforce.
   b. a necessary evil.
   c. an unpleasant reality.
   d. an immoral institution.
   e. superior to adult labor.

81. The author implies that *the catalyst* (line 26) behind the spread of American textile mills in the early 1800s was
   a. Slater's invention of a water-powered spinning machine.
   b. the decline in the ideal of the self-sufficient American farm family.
   c. the expertise of the workforce trained in Slater's prototype mill.
   d. an increased willingness to employ child laborers.
   e. the support of British manufacturers who owned stock in American mills.

82. In line 29, *modeled* most nearly means
   a. posed.
   b. displayed.
   c. arranged.
   d. illustrated.
   e. fashioned.

83. Which of the following techniques is used in the last paragraph of the passage (lines 40–49)?
   a. explanation of terms
   b. description of consensus reached by historians
   c. contrast of different viewpoints
   d. generalized statement
   e. illustration by example

Question 84–91 are based on the following passage.

The following passage describes the Great Depression and the relief policies introduced under President Franklin Delano Roosevelt that aimed to mitigate the effects of the crisis.

(1)   The worst and longest economic crisis in the modern industrial world, the Great Depression in the United States had devastating consequences for American society. At its lowest depth (1932–33), more than 16 million people were unemployed, more than 5,000 banks had
(5)   closed, and over 85,000 businesses had failed. Millions of Americans lost their jobs, their savings, and even their homes. The homeless built shacks for temporary shelter—these emerging shantytowns were nicknamed "Hoovervilles," a bitter homage to President Herbert Hoover, who refused to give government assistance to the jobless. Farmers
(10)   were hit especially hard. A severe drought coupled with the economic crisis ruined small farms throughout the Great Plains as productive

farmland turned to dust and crop prices dropped by 50%. The effects of the American depression—severe unemployment rates and a sharp drop in the production and sales of goods—could also be felt abroad,
(15) where many European nations were still struggling to recover from World War I.

Although the stock market crash of 1929 marked the onset of the depression, it was not *the* cause of it: deep underlying fissures already existed in the economy of America's Roaring Twenties. For example,
(20) the tariff and war-debt policies after World War I contributed to the instability of the banking system. American banks made loans to European countries following World War I. However, the United States kept high tariffs on goods imported from other nations. These policies worked against one another: If other countries could not sell
(25) goods in the United States, they could not make enough money to pay back their loans or to buy American goods.

And while the United States seemed to be enjoying a prosperous period in the 1920s, the wealth was not evenly distributed. Businesses made gains in productivity, but only one segment of the population—
(30) the wealthy—reaped large profits. Workers received only a small share of the wealth they helped produce. At the same time, Americans spent more than they earned. Advertising encouraged Americans to buy cars, radios, and household appliances instead of saving or purchasing only what they could afford. Easy credit polices allowed consumers to
(35) borrow money and accumulate debt. Investors also wildly speculated on the stock market, often borrowing money on credit to buy shares of a company. Stocks increased beyond their worth, but investors were willing to pay inflated prices because they believed stocks would continue to rise. This bubble burst in the fall of 1929, when investors lost
(40) confidence that stock prices would keep rising. As investors sold off stocks, the market spiraled downward. The stock market crash affected the economy in the same way that a stressful event can affect the human body, lowering its resistance to infection.

The ensuing depression led to the election of President Franklin D.
(45) Roosevelt in 1932. Roosevelt introduced relief measures that would revive the economy and bring needed relief to Americans who were suffering the effects of the depression. In his first hundred days in office, Roosevelt and Congress passed major legislation that saved banks from closing and regained public confidence. These measures,
(50) called the New Deal, included the Agricultural Adjustment Act, which paid farmers to slow their production in order to stabilize food prices; the Federal Deposit Insurance Corporation, which insured bank deposits in the case that banks fail; and the Securities and Exchange

Commission, which regulated the stock market. Although the New
(55) Deal offered relief, it did not end the depression. The economy sagged
until the nation entered World War II. However, the New Deal
changed the relationship between government and American citizens,
by expanding the role of the central government in regulating the
economy and creating social assistance programs.

84. The author's main point about the Great Depression is that
  a. government policies had nothing to do with it.
  b. the government immediately stepped in with assistance for the
    jobless and homeless.
  c. underlying problems in the economy preceded it.
  d. the New Deal policies introduced by Franklin D. Roosevelt
    ended it.
  e. its effects were severe but not far-reaching.

85. The passage is best described as
  a. an account of the causes and effects of a major event.
  b. a statement supporting the value of federal social policies.
  c. a condemnation of outdated beliefs.
  d. a polite response to a controversial issue.
  e. a comparison of economic conditions in the 1930s and that of
    today.

86. The author cites the emergence of *"Hoovervilles"* (line 8) as an
  example of
  a. federally sponsored housing programs.
  b. the resilience of Americans who lost their jobs, savings,
    and homes.
  c. the government's unwillingness to assist citizens in desperate
    circumstances.
  d. a new paradigm of "safety net" social programs introduced by
    the government.
  e. the effectiveness of the Hoover administration in dealing with
    the crisis.

87. In line 10, *coupled* most nearly means
  a. eloped.
  b. allied.
  c. centralized.
  d. combined.
  e. associated.

88. The term *policies* as it is used in lines 23–24 most nearly means
  a. theories.
  b. practices.
  c. laws.
  d. examples.
  e. problems.

89. The passage suggests that the 1920s was a decade that extolled the value of
  a. thrift.
  b. prudence.
  c. balance.
  d. tranquility.
  e. extravagance.

90. The example of the human body as a metaphor for the economy (lines 41–43) suggests that
  a. a stressful event like the stock market crash of 1929 probably made a lot of people sick.
  b. the crash weakened the economy's ability to withstand other pressures.
  c. the crash was an untreatable disease.
  d. a single event caused the collapse of the economy.
  e. there is no way to "diagnose" the factors that led to the depression.

91. The content of the last paragraph of the passage (lines 44–59) would most likely support which of the following statements?
  a. The New Deal policies were not radical enough in challenging capitalism.
  b. The economic policies of the New Deal brought about a complete business recovery.
  c. The Agricultural Adjustment Act paid farmers to produce surplus crops.
  d. The federal government became more involved in caring for needy members of society.
  e. The New Deal measures went too far in turning the country toward socialism.

Questions 92–101 are based on the following passage.

In 1804 President Thomas Jefferson sent Army Officers Meriwether Lewis and William Clark on an expedition to explore the territory of the Louisiana Purchase and beyond and to look for a waterway that would connect the Atlantic and Pacific Oceans. This passage describes the collision of cultures that occurred between Native Americans and the representatives of the United States government.

(1)   When Thomas Jefferson sent Lewis and Clark into the West, he patterned their mission on the methods of Enlightenment science: to observe, collect, document, and classify. Such strategies were already in place for the epic voyages made by explorers like Cook and Van-

(5)   couver. Like their contemporaries, Lewis and Clark were more than representatives of European rationalism. They also represented a rising American empire, one built on aggressive territorial expansion and commercial gain.

    But there was another view of the West: that of the native inhabi-

(10)   tants of the land. Their understandings of landscapes, peoples, and resources formed both a contrast and counterpoint to those of Jefferson's travelers. One of Lewis and Clark's missions was to open diplomatic relations between the United States and the Native American nations of the West. As Jefferson told Lewis, "it will now be proper

(15)   you should inform those through whose country you will pass . . . that henceforth we become their fathers and friends." When Euro-Americans and Native Americans met, they used ancient diplomatic protocols that included formal language, ceremonial gifts, and displays of military power. But behind these symbols and rituals there were often

(20)   very different ways of understanding power and authority. Such differences sometimes made communication across the cultural divide difficult and open to confusion and misunderstanding.

    An important organizing principle in Euro-American society was hierarchy. Both soldiers and civilians had complex gradations of rank

(25)   to define who gave orders and who obeyed. While kinship was important in the Euro-American world, it was even more fundamental in tribal societies. Everyone's power and place depended on a complex network of real and symbolic relationships. When the two groups met—whether for trade or diplomacy—each tried to reshape the other

(30)   in their own image. Lewis and Clark sought to impose their own notions of hierarchy on Native Americans by "making chiefs" with medals, printed certificates, and gifts. Native people tried to impose the obligations of kinship on the visitors by means of adoption ceremonies, shared names, and ritual gifts.

(35)   The American republic began to issue peace medals during the first Washington administration, continuing a tradition established by the European nations. Lewis and Clark brought at least eighty-nine medals in five sizes in order to designate five "ranks" of chief. In the eyes of Americans, Native Americans who accepted such medals were
(40) also acknowledging American sovereignty as "children" of a new "great father." And in a moment of imperial bravado, Lewis hung a peace medal around the neck of a Piegan Blackfeet warrior killed by the expedition in late July 1806. As Lewis later explained, he used a peace medal as a way to let the Blackfeet know "who we were."

(45)   In tribal society, kinship was like a legal system—people depended on relatives to protect them from crime, war, and misfortune. People with no kin were outside of society and its rules. To adopt Lewis and Clark into tribal society, the Plains Indians used a pipe ceremony. The ritual of smoking and sharing the pipe was at the heart of much Native
(50) American diplomacy. With the pipe the captains accepted sacred obligations to share wealth, aid in war, and revenge injustice. At the end of the ceremony, the pipe was presented to them so they would never forget their obligations.

     Gift giving was an essential part of diplomacy. To Native Americans, gifts proved the giver's sincerity and honored the tribe. To Lewis
(55) and Clark, some gifts advertised the technological superiority and others encouraged the Native Americans to adopt an agrarian lifestyle. Like salesmen handing out free samples, Lewis and Clark packed bales of manufactured goods to open diplomatic relations with Native
(60) American tribes. Jefferson advised Lewis to give out corn mills to introduce the Native Americans to mechanized agriculture as part of his plan to "civilize and instruct" them. Clark believed the mills were "verry Thankfully recived," but by the next year the Mandan had demolished theirs to use the metal for weapons.

92. The goals of the Lewis and Clark expedition include all of the following purposes EXCEPT to
   a. expand scientific knowledge.
   b. strengthen American claims to western territory.
   c. overcome Native American resistance with military force.
   d. introduce native inhabitants to the ways of Euro-American culture.
   e. make peaceful contact with native inhabitants.

93. According to the passage, the United States government primarily viewed its role in relation to Native Americans as one of
   a. creator.
   b. master.
   c. admirer.
   d. collaborator.
   e. agitator.

94. The word *protocols* as it is used in line 17 most nearly means
   a. beliefs.
   b. tenets.
   c. codes.
   d. tactics.
   e. endeavors.

95. According to the passage, the distribution of peace medals exemplifies
   a. the American republic's attempt to forge a relationship of equals with native people.
   b. a cultural bridge connecting the Euro-Americans with Native American tribes.
   c. the explorers' respect for Native American sovereignty.
   d. the imposition of societal hierarchy on Native Americans.
   e. the acknowledgment of the power and authority of Native American chiefs.

96. The description of Lewis' actions in lines 41–43 is used to
   a. depict the expedition in a patriotic light.
   b. contradict commonly held views of imperialism.
   c. make an ironic statement about the meaning of the peace medals.
   d. give an explanation for the killing of a Piegan Blackfeet warrior.
   e. provide a balanced report of two opposing points of view.

97. The description of the pipe ceremony in lines 48-53 is used to illustrate
   a. the naiveté of the Plains Native Americans.
   b. cultural confusion.
   c. the superiority of the native inhabitants.
   d. how Plains Native Americans honored low-ranking members of society.
   e. the addictive properties of tobacco.

98. In line 47, *adopt* most nearly means
  a. advocate.
  b. nurture.
  c. promote.
  d. foster.
  e. practice.

99. The author uses the image of *salesmen handing out free samples* (lines 57–58) in order to
  a. depict Lewis and Clark as entrepreneurs.
  b. illustrate the generosity Lewis and Clark showed the tribal people they met.
  c. suggest that Lewis and Clark hoped to personally profit from their travels.
  d. imply that everyone likes to get something for free.
  e. show the promotional intent behind the explorers' gift-giving.

100. The passage is developed primarily through
  a. the contrast of different abstract principles.
  b. quotations from one specific text.
  c. the analysis of one extended example.
  d. first-person narratives.
  e. recurring symbols.

101. The author's primary purpose in the passage is to
  a. describe Lewis and Clark's expedition into the West.
  b. show the clashing views of the Indian nations versus those of the American republic.
  c. explore the tribal system of kinship.
  d. make an argument supporting Jefferson's quest for scientific knowledge.
  e. criticize Lewis and Clark's use of peace medals to designate the rank of a chief.

Questions 102–112 are based the following passages.

These passages concern themselves with the nineteenth-century arguments made for and against women's right to vote in the United States. Passage 1 is an excerpt from an address by Isabella Beecher Hooker before the International Council of Women in 1888. Passage 2 is an excerpt from an 1878 report from the Senate's Committee on Privileges and Elections in response to a proposed constitutional amendment that would give women the right to vote.

PASSAGE 1

(1)   First let me speak of the constitution of the United States, and assert that there is not a line in it, nor a word, forbidding women to vote; but, properly interpreted, that is, interpreted by the Declaration of Independence, and by the assertions of the Fathers, it actually guarantees

(5)   to women the right to vote in all elections, both state and national. Listen to the preamble to the constitution, and the preamble you know, is the key to what follows; it is the concrete, general statement of the great principles which subsequent articles express in detail. The preamble says: "We, The People of the United States, in order to form

(10)   a more perfect union, establish justice, insure domestic tranquility, provide for the common defense, promote the general welfare, and secure the blessings of liberty to ourselves and our posterity, do ordain and establish this Constitution for the United States of America."

         Commit this to memory, friends; learn it by heart as well as by head,

(15)   and I should have no need to argue the question before you of my right to vote. For women are "people" surely, and desire, as much as men, to say the least, to establish justice and to insure domestic tranquility; and, brothers, you will never insure domestic tranquility in the days to come unless you allow women to vote, who pay taxes and bear

(20)   equally with yourselves all the burdens of society; for they do not mean any longer to submit patiently and quietly to such injustice, and the sooner men understand this and graciously submit to become the political equals of their mothers, wives, and daughters—aye, of their grandmothers, for that is my category, instead of their political mas-

(25)   ters, as they now are, the sooner will this precious domestic tranquility be insured. Women are surely "people," I said, and were when these words were written, and were as anxious as men to establish justice and promote the general welfare, and no one will have the hardihood to deny that our foremothers (have we not talked about our

(30)   forefathers alone long enough?) did their full share in the work of

establishing justice, providing for the common defense, and promoting the general welfare in all those early days.

(35) The truth is, friends, that when liberties had to be gained by the sword and protected by the sword, men necessarily came to the front and seemed to be the only creators and defenders of these liberties; hence all the way down women have been content to do their patriotic work silently and through men, who are the fighters by nature rather than themselves, until the present day; but now at last, when it is established that ballots instead of bullets are to rule the world . . .

(40) now, it is high time that women ceased to attempt to establish justice and promote the general welfare, and secure the blessings of liberty to themselves and their posterity, through the votes of men . . .

PASSAGE 2

(1) This proposed amendment forbids the United States or any State to deny or abridge the right to vote on account of sex. If adopted, it will make several millions of female voters, totally inexperienced in political affairs, quite generally dependent upon the other sex, all incapable

(5) of performing military duty and without the power to enforce the laws which their numerical strength may enable them to make, and comparatively very few of whom wish to assume the irksome and responsible political duties which this measure thrusts upon them.

An experiment so novel, a change so great, should only be made

(10) slowly and in response to a general public demand, of the existence of which there is no evidence before your committee. Petitions from various parts of the country, containing by estimate about 30,000 names, have been presented to Congress asking for this legislation. They were procured through the efforts of woman-suffrage societies, thoroughly

(15) organized, with active and zealous managers. The ease with which signatures may be procured to any petition is well known. The small number of petitioners, when compared with that of the intelligent women in the country, is striking evidence that there exists among them no general desire to take up the heavy burden of governing, which so

(20) many men seek to evade. It would be unjust, unwise, and impolitic to impose that burden on the great mass of women throughout the country who do not wish for it, to gratify the comparatively few who do.

It has been strongly urged that without the right of suffrage women are and will be subjected to great oppression and injustice. But every

(25) one who has examined the subject at all knows that without female suffrage, legislation for years has improved and is still improving the condition of women. The disabilities imposed upon her by the common

law have, one by one, been swept away until in most of the States she
has the full right to her property and all, or nearly all the rights which
(30) can be granted without impairing or destroying the marriage relation.
These changes have been wrought by the spirit of the age, and are not,
generally at least, the result of any agitation by women in their own
behalf.

Nor can women justly complain of any partiality in the adminis-
(35) tration of justice. They have the sympathy of judges and particularly
of juries to an extent which would warrant loud complaint on the part
of their adversaries of the sterner sex. Their appeals to legislatures
against injustice are never unheeded, and there is no doubt that when
any considerable part of the women of any State really wish for the
(40) right to vote it will be granted without the intervention of Congress.

Any State may grant the right of suffrage to women. Some of them
have done so to a limited extent, and perhaps with good results. It is
evident that in some States public opinion is much more strongly in
favor of it than it is in others. Your committee regards it as unwise and
(45) inexpedient to enable three-fourths in number of the States, through
an amendment to the National Constitution, to force woman suffrage
upon the other fourth in which the public opinion of both sexes may
be strongly adverse to such a change.

For these reasons, your committee reports back said resolution with
(50) a recommendation that it be indefinitely postponed.

102. The author of Passage 1 supports her argument by
    a. providing information about the educational levels achieved
       by women.
    b. sharing anecdotes about women who fought in the American
       Revolution.
    c. referring to principles already accepted by her audience.
    d. describing her personal experience as a citizen of the
       United States.
    e. listing the states in the union that had granted women
       voting rights.

103. The phrase *learn it by heart as well as by head* in line Passage 1, line
14 suggests
    a. an emotional and intellectual response.
    b. rote memorization.
    c. learning from experience rather than books.
    d. accepting an argument on faith.
    e. presupposition of an outcome.

104. In line 27 of Passage 1, *anxious* most nearly means
    a. irritable.
    b. neurotic.
    c. apprehensive.
    d. hasty.
    e. eager.

105. Lines 26–32 of Passage 1 portray American women as
    a. rebellious.
    b. ambitious.
    c. patriotic.
    d. uneducated.
    e. vulnerable.

106. Which of the following best describes the author's strategy in Passage 2?
    a. summarizing public perceptions of the issue
    b. anticipating opposing viewpoints and then refuting them
    c. relating an incident and describing its significance
    d. persuading his audience through emotional appeal
    e. providing evidence that supports both sides of the issue

107. As used in Passage 2, line 9, *novel* most nearly means
    a. rare.
    b. original.
    c. untried.
    d. brilliant.
    e. intellectual.

108. In the third paragraph of Passage 2 (lines 23–33), the author characterizes the activists of the women's suffrage movement as
    a. ardent.
    b. courageous.
    c. conformist.
    d. modest.
    e. genteel.

109. The author of Passage 2 cites the example of a woman's right to
her property (lines 29 and 30) in order to

a. show that women are well represented by the legislature even if
they cannot vote.

b. demonstrate that if women can be responsible for property, they
can be responsible voters.

c. prove that unjust laws affect the condition of women.

d. support the belief that political change should happen quickly.

e. argue that political equality strengthens marriages.

110. Which aspect of the topic of women's voting rights is emphasized
in Passage 2, but not in Passage 1?

a. the interpretation of the Constitution

b. the contributions of American women

c. the tax-paying status of women

d. how the judiciary treats women

e. how ready the country is to allow women the right to vote

111. The two authors would most likely agree with which statement?

a. Most women do not desire the right to vote.

b. Women are not meant to be soldiers.

c. Voting is more of a burden than a privilege.

d. American society is ready for female voters.

e. Men and women should be political equals.

112. The approaches of the two passages to the topic differ in that only
Passage 1

a. describes an incident from the author's personal experience.

b. gives a point and argues its counterpoint.

c. cites several specific examples of laws that benefit women.

d. addresses its audience in the second person.

e. recommends an action to be taken.

Answers

48. **a.** The fact that judicial review can override decisions made by the
legislative and executive branches implies that it gives the court
great authority.

49. **c.** To maintain the "life" of the Constitution, the court applies *its
broad provisions to complex new situations* (line 11) that arise in current law.

50. **c.** To *declare* means to make known formally or officially. To *proclaim* is its synonym, with a slightly different connotation. It implies declaring clearly, forcefully, and authoritatively.

51. **e.** The last sentence offers a view in opposition to the points made earlier in the passage supporting the Supreme Court's power to interpret the Constitution.

52. **c.** According to the passage, the Second Confiscation Act passed by Congress in 1862 *provided the desired signal* (line 7), encouraging him to pursue his plan of a proclamation.

53. **b.** The *speechless* (line 14) reaction of Secretaries Seward and Welles implies that they were surprised by the plan and were concerned about its political and military consequences.

54. **e.** One meaning of *qualified* is fitted by training or experience for a given purpose ("he is qualified for the job"). Another meaning is having complied with specific requirements ("she qualified for the marathon"). In this context, *qualified* means limited or modified in some way.

55. **a.** The author calls the Emancipation Proclamation the *crowning achievement* (line 32) of Lincoln's administration.

56. **a.** Lines 14–15 state that *political cartoons can serve as a vehicle for swaying public opinion and can contribute to reform.*

57. **e.** The consonance in the string of verbs *provoke, poke, and persuade* in line 3, as well as the verb choice *skewering* in line 4 expresses a playfulness of tone. The author's description of the cartoon images of Bill Clinton and George W. Bush (lines 9–12) also mirrors the playfulness of the art of caricature.

58. **e.** One meaning of *vehicle* is a way of carrying or transporting something. In this context, *vehicle* refers to a medium, or the means by which an idea is expressed.

59. **d.** The author cites Thomas Nast's symbols for Tammany Hall and the Democratic and Republican Parties as examples of images that have entered the public consciousness and are *still in currency today* (line 19).

60. **b.** The passage illustrates several protest strategies used in the civil rights movement. Choices **c** and **e** are true statements but are too specific to be the primary focus of the passage. Choices **a** and **d** are not described in detail in the passage.

61. **c.** The passage states that Rosa Park's actions and arrest *set off a train of events that generated a momentum the civil rights movement had never before experienced* (lines 10–12).

62. **e.** One meaning of to *test* is to apply a test as a means of analysis or diagnosis. In this context, *test* refers to putting something to a test or challenging something.

63. **d.** The protest at the Greensboro Woolworth lunch counter inspired others. Lines 25–27 state *two weeks later similar demonstrations had spread to several cities, within a year similar peaceful demonstrations took place in over a hundred cities North and South.*

64. **b.** The passage implies that the 1963 March on Washington was a very successful demonstration: it attracted *more than twice* the number (line 35) of people than organizers expected and *riveted the nation's attention* (lines 33–34), drawing attention to the issues that the march promoted.

65. **c.** One meaning of *refrain* is a regularly recurring verse in a song. In this context, *refrain* refers to the recurring phrase "I have a dream," that Reverend Martin Luther King, Jr. used in his famous speech.

66. **d.** The term *second-class citizen* is not a legal state of citizenship, rather it is a descriptive term that refers to a condition in which citizens of a nation are denied the rights and privileges that other citizens enjoy.

67. **e.** The passage does not speculate about the future nor does it describe the racial discrimination that occurs today in the United States.

68. **b.** The first paragraph introduces the passage's thesis and gives an overview about who emigrated to California and why they came.

69. **b.** The passage provides a historical overview supported by facts and interpreted by the author. The author's opinion is evidenced in the last sentence of the passage (lines 48–49): *a testimony to their outstanding achievements and contributions.*

70. **c.** Line 15 states that the *Chinese immigrants proved to be productive and resourceful.* Lines 46 and 47 praises their *speed, dexterity, and outright perseverance.*

71. **e.** The passage states that at the time, the U.S. Constitution *reserved the right of naturalization for white immigrants,* excluding Chinese immigrants. Chinese immigrants could become citizens, depending on the *whim* (line 25) of local governments.

72. **c.** *Enterprise* means an undertaking that is especially risky. It could also mean a unit of economic organization. In this instance, *industry* fits best within the context.

73. **d.** Chinese immigrants faced discriminatory laws that made them *unable to own land or file mining claims* (lines 27 and 28).

74. **a.** One meaning of *reclaim* is to reform or protest improper conduct. Other meanings are to rescue from an undesirable state, or to make something available for human use—this definition applies to the context.

75. **a.** The last sentence provides an example (Chinese immigrants performing hazardous railroad work in brutal conditions) that supports the general thesis of the passage—that Chinese immigrants made major *contributions to opening up the West* (line 49).

76. d. The passage describes the introduction of *British cotton technology to America* (line 15), specifically to New England.

77. b. The passage mentions the Houses of Industry in Boston and Philadelphia (line 5) as an example of the association of cloth manufacturing with *relief of the poor* (lines 6–7).

78. b. The *mounting conflict between the colonies and England* described in line 1 suggests that America had political and/or economic reasons for developing its own textile industry.

79. a. The description of Samuel Slater's immigration to America shows the deceptive measures necessary to evade British export laws and introduce cotton technology to the colonies. Slater posed as a farmer in order to emigrate to America and *committed to memory* (line 20) the cotton technology he learned in English factory.

80. a. The author does not offer Slater's personal viewpoint on child labor, only the fact that Slater hired *nine children between the ages of seven and twelve* (line 23) to work in his Rhode Island mill.

81. c. According to the passage, the knowledge and training acquired in Slater's mill of a *generation of millwrights and textile workers* (line 25) provided the catalyst for the spread of cotton mills in New England.

82. e. One meaning of to *model* is to display by means of wearing, using, or posing. In this context, to *model* means to construct or fashion after a pattern.

83. c. The author offers a contrast of different viewpoints exemplified by the philosophy of Thomas Jefferson, who supported a republic *whose heart was the independent, democratic farmer* (lines 43–44) and that of Alexander Hamilton, who *promoted manufacturing* (line 46) and industrial development.

84. c. According to the passage, *deep underlying fissures* that *already existed in the economy* (lines 18–19) led to the Great Depression.

85. a. The passage is primarily an account that describes the causative factors (for example, tariff and war-debt policies, disproportionate wealth, and the accumulation of debt) that led to the depression and its effects (for example, business failures, bank closings, homelessness, federal relief programs).

86. c. Lines 7–8 state that shantytowns were called "*Hoovervilles*" because citizens blamed their plight on the Hoover administration's refusal to offer assistance. Choice **b** may be true, but the passage does not directly support this claim.

87. d. In this context, *coupled* means to join for combined effect.

88. b. Although policies can refer to regulations or *laws* (choice **c**) or guiding principles or *theories* (choice **a**), in this context *policy* refers to a course or method of action of a government or business intended to influence decisions or actions. Choice **b** is the only selection that implies action.

89. **e.** The passage describes the decade as one in which spending won out over prudent measures like saving (lines 31–32). The wild stock market speculation described in lines 35–37 is another example of the exuberant decade.

90. **b.** The analogy depicts the stock market crash of 1929 as a weakening agent to the economy (the way a stressful event may weaken the body's resistance to illness), not as the sole cause of the depression.

91. **d.** Lines 56–59 state that the New Deal expanded *the role of the central government in regulating the economy and creating social assistance programs.* Choices **b** and **c** are incorrect and choices **a** and **e** require an opinion; the author does not offer his or her viewpoint about the New Deal measures.

92. **c.** The Lewis and Clark expedition did not have a military goal and did not have any violent encounters except the one described in lines 41–43.

93. **b.** Jefferson and his representatives wanted Native Americans to acknowledge American sovereignty and to see themselves as children to his role as their "father."

94. **c.** One meaning of *protocol* is a code that demands strict adherence to etiquette.

95. **d.** The passage states that *Lewis and Clark sought to impose their own notions of hierarchy on Native Americans by "making chiefs" with medals, printed certificates, and gifts* (lines 30–33).

96. **c.** By placing a peace medal around the neck of a man killed by the expedition makes an ironic statement about the meaning of "peace."

97. **b.** To the Plains Native Americans, the pipe ceremony meant that those who participated *accepted sacred obligations to share wealth, aid in war, and revenge injustice* (lines 50–51). The passage suggests that Lewis and Clark most likely did not understand the significance of the ceremony.

98. **e.** One meaning of *adopt* is to take by choice into a relationship. In this context, *adopt* has another meaning: to take up and practice or use.

99. **e.** By giving manufactured goods to Native Americans, Lewis and Clark were promoting Euro-American culture. Jefferson hoped that these *free samples* would *introduce the Native Americans to mechanized agriculture as part of his plan to "civilize and instruct" them* (lines 58–61).

100. **a.** The passage compares different abstract principles, or organizing principles of Euro-American society versus that of tribal societies. For example, it explores the principles of hierarchy and kinship.

101. **b.** Choice **a** is too general to be the primary purpose of the passage, whereas choices **c** and **e** are too specific. Choice **d** is not supported by the passage.

102. **c.** Beecher Hooker invokes the Constitution (line 1) and recites the preamble (lines 9–13) in order to appeal to and persuade her audience.

103. **a.** Beecher Hooker plays on the two meanings suggested by the phrase *learn it by heart as well as by head.* She asks her audience to not only memorize the Constitution's preamble, but to use both emotion and intellect to understand its meaning.

104. **e.** One meaning of *anxious* is extreme uneasiness or dread. An alternative meaning applies to this context—that of ardently or earnestly wishing.

105. **c.** Passage 1 argues that the foremothers of the nation were patriotic and *did their full share* (line 30) of contributing to the early republic.

106. **b.** The passage anticipates the arguments of those in favor of women's right to vote and refutes them.

107. **c.** *Novel* means new and not resembling something known or used in the past. Choice **b,** original, could fit this definition but its connotation is too positive for the context.

108. **a.** Passage 2 describes *woman-suffrage societies* as *thoroughly organized, with active and zealous managers* (lines 14–15). Choice **b,** *courageous,* is too positive for the context of the passage.

109. **a.** Passage 2 states that *every one . . . knows that without female suffrage, legislation for years has improved and is still improving the condition of women* (lines 24–27).

110. **d.** Passage 2 emphasizes how well women are served by judges in line 35. Passage 1 does not refer to this issue at all.

111. **b.** Passage 1 describes men as *fighters by nature* (line 37), but not women. Passage 2 describes women as *incapable of performing military duty* (lines 4–5).

112. **d.** Passage 1 addresses its audience in the second person, whereas Passage 2 does not. Passage 1 also refers to its audience as *friends* (line 14) and *brothers* (line 18).

Arts and Humanities

Questions 113–116 are based on the following passage.

The following paragraph details the design of New York City's Central Park.

(1)   Although it is called Central Park, New York City's great green space has no "center"—no formal walkway down the middle of the park, no central monument or body of water, no single orienting feature. The paths wind, the landscape constantly shifts and changes, the sections

(5)   spill into one another in a seemingly random manner. But this "decentering" was precisely the intent of the park's innovative design. Made to look as natural as possible, Frederick Law Olmsted's 1858 plan for Central Park had as its main goal the creation of a democratic playground— a place with many centers to reflect the multiplicity of its uses and users.

(10)   Olmsted designed the park to allow interaction among the various members of society, without giving preference to one group or class. Thus, Olmsted's ideal of a "commonplace civilization" could be realized.

113. In lines 3–5, the author describes specific park features in order to
    a. present both sides of an argument.
    b. suggest the organization of the rest of the passage.
    c. provide evidence that the park has no center.
    d. demonstrate how large the park is.
    e. show how well the author knows the park.

114. The main idea of this passage is that
   a. New York City is a democratic city.
   b. Olmsted was a brilliant designer.
   c. More parks should be designed without centers.
   d. Central Park is used by many people for many different purposes.
   e. Central Park is democratic by design.

115. The passage suggests that Olmsted's design
   a. was like most other parks being designed at the time.
   b. was radically different from other park designs.
   c. was initially very unpopular with New Yorkers.
   d. was inspired by similar parks in Europe.
   e. did not succeed in creating a democratic playground.

116. The word *commonplace* as used in line 12 most nearly means
   a. inclusive.
   b. ordinary.
   c. mediocre.
   d. normal.
   e. trite.

Questions 117–120 are based on the following passage.

In this excerpt from Book One of his Nicomachean Ethics, *Aristotle expands his definitions of "good" and "happiness."*

(1)   Good things are commonly divided into three classes: (1) external goods, (2) goods of the soul, and (3) goods of the body. Of these, we call the goods pertaining to the soul goods in the highest and fullest sense. But in speaking of "soul," we refer to our soul's actions and
(5)   activities. Thus, our definition [of good] tallies with this opinion which has been current for a long time and to which philosophers subscribe. We are also right in defining the end as consisting of actions and activities; for in this way the end is included among the goods of the soul and not among external goods.
(10)   Also the view that a happy man lives well and fares well fits in with our definition: for we have all but defined happiness as a kind of good life and well-being.
         Moreover, the characteristics which one looks for in happiness are all included in our definition. For some people think that happiness is
(15)   a virtue, others that it is practical wisdom, others that it is some kind

of theoretical wisdom; others again believe it to be all or some of these accompanied by, or not devoid of, pleasure; and some people also include external prosperity in its definition.

117. According to the passage, the greatest goods are those that
  a. are theoretical.
  b. are spiritual.
  c. are intellectual.
  d. create happiness.
  e. create prosperity.

118. The word *tallies* in line 5 means
  a. keeps count.
  b. records.
  c. labels.
  d. corresponds.
  e. scores.

119. The author's definition of happiness in lines 11–12 is related to the definition of good in that
  a. living a good life will bring you happiness.
  b. happiness is the same as goodness.
  c. happiness is often sacrificed to attain the good.
  d. all things that create happiness are good things.
  e. happiness is a virtue.

120. In lines 13–18, the author's main purpose is to
  a. show that different people have different definitions of happiness.
  b. define virtue.
  c. prove that his definition of happiness is valid.
  d. explain the relationship between happiness and goodness.
  e. provide guidelines for good behavior.

Questions 121–125 are based on the following passage.

The following passage describes the ethical theory of utilitarianism.

(1)   If you have ever made a list of pros and cons to help you make a decision, you have used the utilitarian method of moral reasoning. One of the main ethical theories, utilitarianism posits that the key to deciding what makes an act morally right or wrong is its consequences.

(5)   Whether our intentions are good or bad is irrelevant; what matters is whether the *result* of our actions is good or bad. To utilitarians, happiness is the ultimate goal of human beings and the highest moral good. Thus, if there is great unhappiness because of an act, then that action can be said to be morally wrong. If, on the other hand, there is

(10)   great happiness because of an action, then that act can be said to be morally right.

      Utilitarians believe that we should carefully weigh the potential consequences of an action before we take it. Will the act lead to things that will make us, or others, happy? Will it make us, or others,

(15)   unhappy? According to utilitarians, we should choose to do that which creates the greatest amount of good (happiness) for the greatest number of people. This can be difficult to determine, though, because sometimes an act can create short-term happiness but misery in the long term. Another problematic aspect of utilitarianism is that it

(20)   deems it acceptable—indeed, even necessary—to use another person as a means to an end and sacrifice the happiness of one or a few for the happiness of many.

121. In lines 1–2, the author refers to a list of pros and cons in order to

    a. show that there are both positive and negative aspects of utilitarianism.

    b. suggest that making a list of pros and cons is not an effective way to make a decision.

    c. emphasize that utilitarians consider both the good and the bad before making a decision.

    d. indicate that readers will learn how to make decisions using pro/con lists.

    e. show readers that they are probably already familiar with the principles of utilitarian reasoning.

122. The word *posits* in line 3 means

    a. agrees.

    b. asserts.

    c. places.

    d. chooses.

    e. denies.

123. According to the definition of utilitarianism in lines 3–11, stealing bread to feed hungry children would be
   a. morally right because it has good intentions.
   b. morally wrong because of it violates another's rights.
   c. morally right because it has positive consequences.
   d. morally wrong because stealing is illegal.
   e. neither morally right nor wrong; a neutral action.

124. According to the utilitarian principles described in lines 13–19, we should
   a. do what will bring us the most happiness.
   b. always think of others first.
   c. make our intentions clear to others.
   d. do what will make the most people the most happy.
   e. avoid things that will make us unhappy.

125. In lines 19–22, the author's purpose is to show that
   a. using utilitarianism to make a moral decision is not always easy.
   b. sacrifice is necessary in life.
   c. long-term consequences are more important than short-term consequences.
   d. a pro/con list is the most effective technique for making an important decision.
   e. great good often comes at a great price.

Questions 126–133 are based on the following passage.

Written by John Henry Newman in 1852, the following passage presents Newman's idea of the purpose and benefits of a university education.

(1)   I have said that all branches of knowledge are connected together, because the subject-matter of knowledge is intimately united in itself [. . .]. Hence it is that the Sciences, into which our knowledge may be said to be cast, have multiple bearings on one another, and an inter-

(5)   nal sympathy, and admit, or rather demand, comparison and adjustment. They complete, correct, and balance each other. This consideration, if well-founded, must be taken into account, not only as regards the attainment of truth, which is their common end, but as regards the influence which they excise upon those whose education

(10)   consists in the study of them. I have already said, that to give undue prominence to one is to be unjust to another; to neglect or supersede these is to divert those from their proper object. It is to unsettle the

boundary lines between science and science, to disturb their action, to destroy the harmony which binds them together. Such a proceeding
(15) will have a corresponding effect when introduced into a place of education. There is no science but tells a different tale, when viewed as a portion of a whole, from what it is likely to suggest when taken by itself, without the safeguard, as I may call it, of others.

Let me make use of an illustration. In the combination of colors,
(20) very different effects are produced by a difference in their selection and juxtaposition; red, green, and white, change their shades, according to the contrast to which they are submitted. And, in like manner, the drift and meaning of a branch of knowledge varies with the company in which it is introduced to the student. If his reading is confined
(25) simply to one subject, however such division of labor may favor the advancement of a particular pursuit, a point into which I do not here enter, certainly it has a tendency to contract his mind. If it is incorporated with others, it depends on those others as to the kind of influence that it exerts upon him. Thus the Classics, which in England are
(30) the means of refining the taste, have in France subserved the spread of revolutionary and deistical doctrines. [. . . .] In a like manner, I suppose, Arcesilas would not have handled logic as Aristotle, nor Aristotle have criticized poets as Plato; yet reasoning and poetry are subject to scientific rules.

(35) It is a great point then to enlarge the range of studies which a University professes, even for the sake of the students; and, though they cannot pursue every subject which is open to them, they will be the gainers by living among those and under those who represent the whole circle. This I conceive to be the advantage of a seat of univer-
(40) sal learning, considered as a place of education. An assemblage of learned men, zealous for their own sciences, and rivals of each other, are brought, by familiar intercourse and for the sake of intellectual peace, to adjust together the claims and relations of their respective subjects of investigation. They learn to respect, to consult, to aid each
(45) other. Thus is created a pure and clear atmosphere of thought, which the student also breathes, though in his own case he only pursues a few sciences out of the multitude. He profits by an intellectual tradition, which is independent of particular teachers, which guides him in his choice of subjects, and duly interprets for him those which he chooses.
(50) He apprehends the great outlines of knowledge, the principles on which it rests, the scale of its parts, its lights and its shades, its great points and its little, as he otherwise cannot apprehend them. Hence it is that his education is called "Liberal." A habit of mind is formed which lasts through life, of which the attributes are, freedom, equi-

(55)   tableness, calmness, moderation, and wisdom; or what in a former dis-
course I have ventured to call a philosophical habit. This then I would
assign as the special fruit of the education furnished at a University, as
contrasted with other places of teaching or modes of teaching. This is
the main purpose of a University in its treatment of its students.

126. The main idea of the first paragraph (lines 1–18) is that
   a. each science should be studied independently.
   b. the sciences are interrelated.
   c. the boundary lines between each of the sciences should be
      clearer.
   d. some sciences are unduly given more emphasis than others at
      the university level.
   e. it is difficult to attain a proper balance among the sciences.

127. By *the Sciences* (line 3), the author means
   a. the physical sciences only.
   b. the social sciences only.
   c. the physical and social sciences.
   d. all branches of knowledge, including the physical and social
      sciences and the humanities.
   e. educational methodologies.

128. The word *excise* in line 9 most nearly means
   a. remove.
   b. cut.
   c. impose.
   d. arrange.
   e. compete.

129. By using the word *safeguard* in line 18, the author suggests that
   a. it is dangerous to limit one's education to one field or area of
      specialization.
   b. it is not safe to study the sciences.
   c. the more one knows, the safer one will feel.
   d. one should choose a second area of specialization as a backup in
      case the first does not work out.
   e. each science has its own specific safety guidelines.

130. The purpose of the second paragraph (lines 19–34) is to
   a. introduce a new idea.
   b. develop the idea presented in the previous paragraph.
   c. state the main idea of the passage.
   d. present an alternative point of view.
   e. compare and contrast different branches of knowledge.

131. The word *apprehends* as used in lines 50 and 52 means
   a. understands.
   b. captures.
   c. fears.
   d. believes.
   e. contains.

132. Which of the following best describes the author's idea of a liberal education?
   a. in-depth specialization in one area.
   b. free education for all.
   c. a broad scope of knowledge in several disciplines.
   d. training for a scientific career.
   e. an emphasis on the arts rather than the sciences.

133. The author believes that a university should
   I. have faculty representing a wide range of subjects and philosophies
   II. teach students how to see the relationships among ideas
   III. teach students to understand and respect other points of view
   IV. teach students liberal rather than conservative ideals
   a. I and II only
   b. I, II, and III
   c. I and IV
   d. IV only
   e. all of the above

Questions 134–141 are based on the following passage.

In this passage, the author discusses the problem of maintaining privacy in our high-tech society.

(1)   A recent *New York Times* "House and Home" article featured the story of a man who lives in a glass house. Every wall in his home is transparent; he has no walls to hide behind, not even in the bathroom. Of

(5) course, he lives in an isolated area, so he doesn't exactly have neighbors peering in and watching his every move. But he has chosen to live without any physical privacy in a home that allows every action to be seen. He has created his own panopticon of sorts, a place in which everything is in full view of others.

(10) The term *panopticon* was coined by Jeremy Bentham in the late eighteenth century when he was describing an idea for how prisons should be designed. The prisoner's cells would be placed in a circle with a guard tower in the middle. All walls facing the center of the circle would be glass. In that way, every prisoner's cell would be in full view of the guards. The prisoners could do nothing unobserved, but *(15)* the prisoners would not be able to see the guard tower. They would know they were being watched—or rather, they would know that they *could* be being watched—but because they could not see the observer, they would never know when the guard was actually monitoring their actions.

(20) It is common knowledge that people behave differently when they know they are being watched. We act differently when we know someone is looking; we act differently when we think someone else *might* be looking. In these situations, we are less likely to be ourselves; instead, we will act the way we think we should act when we are being *(25)* observed by others.

In our wired society, many talk of the panopticon as a metaphor for the future. But in many ways, the panopticon is already here. Surveillance cameras are everywhere, and we often don't even know our actions are being recorded. In fact, the surveillance camera industry is *(30)* enormous, and these cameras keep getting smaller and smaller to make surveillance easier and more ubiquitous. In addition, we leave a record of everything we do online; our cyber-whereabouts can be tracked and that information used for various purposes. Every time we use a credit card, make a major purchase, answer a survey, apply for a *(35)* loan, or join a mailing list, our actions are observed and recorded. And most of us have no idea just how much information about us has been recorded and how much data is available to various sources. The scale of information gathering and the scale of exchange have both expanded so rapidly in the last decade that there are now millions of *(40)* electronic profiles of individuals existing in cyberspace, profiles that are bought and sold, traded, and often used for important decisions, such as whether or not to grant someone a loan. However, that information is essentially beyond our control. We can do little to stop the information gathering and exchange and can only hope to be able to *(45)* control the damage if something goes wrong.

Something went wrong recently for me. Someone obtained my Social Security number, address, work number and address, and a few other vital pieces of data. That person then applied for a credit account in my name. The application was approved, and I soon received a bill
(50)   for nearly $5,000 worth of computer-related purchases.

Fraud, of course, is a different issue, but this kind of fraud couldn't happen—or at least, couldn't happen with such ease and frequency—in a world of paper-based records. With so much information floating about in cyberspace, and so much technology that can record and
(55)   observe, our privacy has been deeply compromised.

I find it truly amazing that someone would want to live in a transparent house at any time, but especially in an age when individual privacy is becoming increasingly difficult to maintain and defend (against those who argue that information must be gathered for the social
(60)   good). Or perhaps this man's house is an attempt to call our attention to the fact that the panopticon is already here, and that we are all just as exposed as he is.

134. According to the passage, a *panopticon* is
   a. a prison cell.
   b. a place in which everything can be seen by others.
   c. a tower that provides a panoramic view.
   d. a house that is transparent.
   e. a place in which surveillance cameras and other monitoring equipment are in use.

135. The description of how the panopticon would work in a prison (lines 10–19) implies that the panopticon
   a. can be an effective tool for social control.
   b. should be used regularly in public places.
   c. is not applicable outside of the prison dynamic.
   d. is an effective tool for sharing information.
   c. will redefine privacy for the twenty-first century.

136. In lines 26–36, the author suggests that the panopticon is a metaphor for our society because
   a. our privacy is transparent.
   b. we are all prisoners in our own homes.
   c. our actions are constantly observed and recorded.
   d. we are always afraid that someone might be watching us.
   e. there is rampant exchange of information in cyberspace.

137. According to the passage, a key difference between the prison panopticon and the modern technological panopticon is that
   a. the prisoners can see their observers, but we can't.
   b. today's prisons are too crowded for the panopticon to work.
   c. prisoners are less informed about privacy issues than technology users.
   d. the prisoners are aware that they may be being watched, but we often don't even know we are being monitored.
   e. prisoners are more protected in their panopticon than we are in ours.

138. The passage suggests that all of the following contribute to the erosion of privacy EXCEPT
   a. increased use of credit cards for purchases.
   b. buying and selling of electronic profiles.
   c. increasingly discreet surveillance equipment.
   d. lack of controls over information exchange.
   e. easy access to electronic information in cyberspace.

139. The author describes a personal experience with identity theft in order to
   a. show how prevalent identity theft is.
   b. show how angry he is about having his privacy invaded.
   c. show an example of how private information can be taken and misused.
   d. demonstrate a flaw in the panopticon.
   e. demonstrate the vast scale of information exchange.

140. The word *compromised* in line 55 means
   a. conceded.
   b. agreed.
   c. dishonored.
   d. negotiated.
   e. jeopardized.

141. Based on the passage, it can be inferred that the author would support which of the following?
   a. widespread construction of glass houses
   b. stricter sentencing for perpetrators of fraud
   c. greater flexibility in loan approval criteria
   d. stricter regulations for information gathering and exchange
   e. modeling prisons after Bentham's panopticon

Questions 142–149 are based on the following passage.

The following passage tells of the mythological Greek god Prometheus.

(1)   Without a doubt, one of the most interesting mythological characters is the Greek god Prometheus. A complex character with an undying love for the human beings he created, Prometheus embodies a rich combination of often contradictory characteristics, including loyalty

(5)   and defiance, trickery and trustworthiness. He shows resilience and resolve in his actions yet weakness in his fondness for humankind.

To reward Prometheus (whose name means "forethought") and his brother Epimetheus ("afterthought") for helping him defeat the Titans, Zeus, the great ruler of Olympian gods, gave the brothers the

(10)   task of creating mortals to populate the land around Mount Olympus. Prometheus asked Epimetheus to give the creatures their various characteristics, such as cunning, swiftness, and flight. By the time he got to man, however, there was nothing left to give. So Prometheus decided to make man in his image: he stood man upright like the gods

(15)   and became the benefactor and protector of mankind.

Though Prometheus was particularly fond of his creation, Zeus didn't care for mankind and didn't want men to have the divine gift of knowledge. But Prometheus took pity on mortal men and gave them knowledge of the arts and sciences, including the healing arts and agri-

(20)   culture.

Always seeking the best for his creation, one day Prometheus conspired to trick Zeus to give the best meat of an ox to men instead of Zeus. He cut up the ox and hid the bones in layers of fat; then he hid the meat and innards inside the hide. When Prometheus presented the

(25)   piles to Zeus, Zeus chose the pile that looked like fat and meat. He was enraged to find that it was nothing but bones.

To punish Prometheus for his deceit and his fondness for humans, Zeus forbade men fire—a symbol of creative power, life force, and divine knowledge. But Prometheus would not let his children be

(30)   denied this greatest of gifts. He took a hollow reed, stole fire from Mount Olympus, and gave it to man. With this divine power, creativity, ingenuity, and culture flourished in the land of mortals.

Again Zeus punished man for Prometheus's transgression, this time by sending the first woman, Pandora, to Earth. Pandora brought with

(35)   her a "gift" from Zeus: a jar filled with evils of every kind. Prometheus knew Zeus to be vengeful and warned Epimetheus not to accept any gifts from Zeus, but Epimetheus was too taken with Pandora's beauty and allowed her to stay. Eventually Pandora opened the jar she'd been forbidden to open, releasing all manner of evils, including Treachery,

(40)   Sorrow, Villainy, Misfortune, and Plague. At the bottom of the jar was
       Hope, but Pandora closed the lid before Hope could escape.
           Prometheus drew Zeus's greatest wrath when he refused to tell Zeus
       which of Zeus's sons would kill him and take over the throne. Believ-
       ing he could torture Prometheus into revealing the secret, Zeus bound
(45)   Prometheus to a rock where every day an eagle would come to tear at
       his flesh and eat his liver, which would regenerate each night. But
       Prometheus refused to reveal his knowledge of the future to Zeus and
       maintained his silence. Eventually, Prometheus was released by Her-
       acles (also known as Hercules), the last mortal son of Zeus and the
(50)   strongest of all mortals. Soon afterwards, Prometheus received
       immortality from a dying centaur, to take his place forever among the
       great gods of Olympus.

   142. The main idea of the first paragraph (lines 1–6) is that Prometheus
       a. is disrespectful of authority.
       b. is the mythological creator of humans.
       c. has many admirable characteristics.
       d. should not have been so fond of humans.
       e. is a fascinating character because of his complexity.

   143. The author's primary purpose in this passage is to
       a. demonstrate the vengeful nature of Zeus.
       b. show how much Prometheus cared for humans.
       c. create in readers an interest in mythology.
       d. relate the story of Prometheus.
       e. prove that Prometheus, not Zeus, was the creator of man.

   144. Based on this passage, it can be inferred that Zeus disliked humans
       because
       a. Prometheus spent too much time with them.
       b. Prometheus cared for humans more than he did for Zeus.
       c. humans could not be trusted.
       d. humans did not respect Zeus.
       e. he did not create them.

   145. Zeus becomes angry at Prometheus for all of the following EXCEPT
       a. creating man.
       b. giving man fire.
       c. being excessively fond of humans.
       d. refusing to reveal which of his sons would kill him.
       e. tricking him into taking the undesirable part of an ox.

146. Based on the passage, the relationship between Prometheus and humans can best be described as that of
   a. parent and child.
   b. close friends.
   c. master and servant.
   d. bitter enemies.
   e. reluctant allies.

147. The word *transgression* as used in line 33 means
   a. villainy.
   b. trespass.
   c. irregularity.
   d. error.
   e. disobedience.

148. The fact that Zeus included Hope in Pandora's jar (lines 38–41) suggests that
   a. Zeus really did love humans as much as Prometheus did.
   b. while Zeus was a vengeful god, he did not wish humans to live in utter despair.
   c. Zeus was just playing a trick on humans.
   d. Zeus was trying to make amends with Prometheus.
   e. Zeus wanted to drive Prometheus away from humans.

149. The content and style of this passage suggest that the intended audience
   a. are experts on Greek mythology.
   b. are religious officials.
   c. is a general lay audience.
   d. are family members and friends.
   e. is a scholarly review board.

Questions 150–158 are based on the following passage.

The following passage describes an influential group of nineteenth century painters.

(1)   When one thinks of student-led rebellions and the changes they can create, one typically thinks of the struggles of the twentieth century, such as the civil rights movement or anti-war protests of the sixties. But there have been less dramatic, though no less passionate, rebel-
(5)   lions led by young activists in previous centuries—rebellions that had

lasting impact on the world around us. One such example is the Pre-Raphaelite Brotherhood.

(10) In the mid-1800s, the art world in England was rattled by the initials PRB. The PRB (or Pre-Raphaelite Brotherhood) was founded by William Holman Hunt, John Everett Millais, and Dante Gabriel Rossetti. These three burgeoning artists (the oldest of whom was 21) and their disdain for the artistic conventions of the time would have a dramatic influence on the art world for generations to come.

(15) The PRB was formed in response to the brotherhood's belief that the current popular art being produced in England was lacking in meaning and aesthetic honesty. During the era leading up to the PRB, the Royal Academy dominated British art. The Royal Academy advocated a style that was typically staid and relied heavily upon the use of dark amber and brown tones to depict overly idealized landscapes,

(20) carefully arranged family portraits and still lifes, and overly dramatic nature scenes such as a boat caught in stormy seas. By contrast, the PRB believed that art should present subjects that, by their very nature, had greater meaning and more accurately depicted reality. The PRB was committed to bringing greater integrity to art and even went

(25) so far as to publish *The Germ*, a journal that extolled the virtues of the PRB's aesthetic principles.

To develop subjects with greater meaning, the PRB initially turned to ancient myths and stories from the Bible. Many of the PRB's biblically themed paintings portrayed the religious figures as regular peo-

(30) ple. This departure from the convention of the time is notable in John Everett Millais' *Christ in the Home of his Parents*. In this painting, Jesus is portrayed as a young boy in his father's carpentry shop. Everyone in the painting, including Christ himself, looks like a common person of that time period, complete with dirty feet and hands. This realism—

(35) especially as it related to the Biblical figures—was not well received by many in the art world at the time. Later works done by fellow PRB members, and those inspired by them, utilized themes from poetry, literature, and medieval tales, often with the aim of highlighting the societal and moral challenges of the time.

(40) With the goal of bringing greater honesty to their work, the PRB ignored the convention of painting an imagined or remembered landscape or background. Instead, PRB members would hunt (sometimes for weeks) for locations to incorporate into their paintings and then paint them in exacting detail.

(45) One of the most distinctive aspects of PRB works—both in contrast to the works produced during the early nineteenth century and with the art of today—is their dramatic use of color. By committing them-

selves to the accurate depiction of nature, the PRB brought a freshness
and drama to its work through the copious use of color. Further
(50) enhancing their work was a technique they used which involved applying the colored paint on top of wet white paint previously applied to
their canvasses. The effect was to make the colors even brighter and
more dramatic. Even today, more than 150 years later, PRB paintings
have a luminescence beyond those of other works from the same time
(55) period. It is believed that their paintings have this quality today
because the white layer underneath the colored paint continues to add
brightness and life to the painting.

Originally founded by three upstart young men, the PRB had a
tremendous influence on an entire generation of artists. William Mor-
(60) ris, Ford Maddox Brown, and Edward Burne-Jones are just a few of
the significant artists of the time whose work was dramatically influenced by the PRB.

150. The word *upstart* in line 58 means
    a. well-regarded.
    b. conceited.
    c. beginning from an advanced position.
    d. suddenly raised to a high position.
    e. receiving numerous honors.

151. In the opening paragraphs (lines 1–7), the author characterizes the
PRB as all of the following EXCEPT
    a. young.
    b. revolutionary.
    c. rebellious.
    d. anti-war.
    e. passionate.

152. The word *burgeoning* in line 11 means
    a. bursting.
    b. developing.
    c. flourishing.
    d. expanding.
    e. prospering.

153. The PRB believed artists should do all of the following EXCEPT
   a. paint meaningful subjects.
   b. paint existing rather than imagined landscapes.
   c. use vibrant colors.
   d. choose subjects that address social issues.
   e. portray people and nature in an idealized manner.

154. According to the passage, the art world
   a. disliked the PRB's emphasis on realism.
   b. disdained the PRB's choice of subject matter.
   c. appreciated the PRB's attention to detail.
   d. embraced the PRB's style, especially their use of color.
   e. was offended by the PRB's attempts to change the Royal Academy's style.

155. The PRB's rebellion was rooted in
   a. a fascination with religious and mythological subjects.
   b. similar artistic rebellions in Europe.
   c. a belief that their peers' work lacked integrity.
   d. a distrust of realistic landscapes and poetic themes.
   e. a conflict over the use of color in painting.

156. According to the author, the most distinguishing feature of PRB works is their
   a. surrealism.
   b. contrast to Royal Academy art.
   c. everyday subject matter.
   d. stoicism.
   e. vibrant colors.

157. The author's main purpose in this passage is to
   a. describe the lives of the founders of the PRB.
   b. describe the artistic principles of the PRB.
   c. compare and contrast revolutions in art.
   d. describe the controversy created by the PRB.
   e. describe how the PRB influenced future artists.

158. It can be inferred that members of the PRB

    a. were more socially conscious than members of the Royal Academy.

    b. were more educated than the members of the Royal Academy.

    c. were more popular than members of the Royal Academy.

    d. were bitter about being excluded from the Royal Academy.

    e. had a great deal of influence within the Royal Academy.

Questions 159–167 are based on the following passage.

In the following passage the author tells of public art and its functions.

(1)   In Manhattan's Eighth Avenue/Fourteenth Street subway station, a grinning bronze alligator with human hands pops out of a manhole cover to grab a bronze "baby" whose head is the shape of a moneybag. In the Bronx General Post Office, a giant 13-panel painting called

(5)   *Resources of America* celebrates the hard work and industrialism of America in the first half of the twentieth century. And in Brooklyn's MetroTech Center just over the Brooklyn Bridge, several installations of art are on view at any given time—from an iron lasso resembling a giant charm bracelet to a series of wagons that play recordings of great

(10)   American poems to a life-sized seeing eye dog that looks so real people are constantly stopping to pet it.

    There exists in every city a symbiotic relationship between the city and its art. When we hear the term *art*, we tend to think of private art—the kind displayed in private spaces such as museums, concert

(15)   halls, and galleries. But there is a growing interest in, and respect for, public art: the kind of art created for and displayed in public spaces such as parks, building lobbies, and sidewalks.

    Although all art is inherently public—created in order to convey an idea or emotion to others—"public art," as opposed to art that is

(20)   sequestered in museums and galleries, is art specifically designed for a public arena where the art will be encountered by people in their normal day-to-day activities. Public art can be purely ornamental or highly functional; it can be as subtle as a decorative door knob or as conspicuous as the Chicago Picasso. It is also an essential element of

(25)   effective urban design.

    The more obvious forms of public art include monuments, sculptures, fountains, murals, and gardens. But public art also takes the form of ornamental benches or street lights, decorative manhole covers, and mosaics on trash bins. Many city dwellers would be surprised

(30)   to discover just how much public art is really around them and how

much art they have passed by without noticing, and how much impact public art has on their day-to-day lives.

(35)

(40)

Public art fulfills several functions essential to the health of a city and its citizens. It educates about history and culture—of the artist, the neighborhood, the city, the nation. Public art is also a "place-making device" that instantly creates memorable, experiential landmarks, fashioning a unique identity for a public place, personalizing it and giving it a specific character. It stimulates the public, challenging viewers to interpret the art and arousing their emotions, and it promotes community by stimulating interaction among viewers. In serving these multiple and important functions, public art beautifies the area and regenerates both the place and the viewer.

(45)

(50)

One question often debated in public art forums is whether public art should be created *with* or *by* the public rather than *for* the public. Increasingly, cities and artists are recognizing the importance of creating works with meaning for the intended audience, and this generally requires direct input from the community or from an artist entrenched in that community. At the same time, however, art created for the community by an "outsider" often adds fresh perspective. Thus, cities and their citizens are best served by a combination of public art created *by* members of the community, art created with input *from* members of the community, and art created by others *for* the community.

159. The primary purpose of the opening paragraph is to
   a. show how entertaining public art can be.
   b. introduce readers to the idea of public art.
   c. define public art.
   d. get readers to pay more attention to public art.
   e. show the prevalence and diversity of public art.

160. The word *inherently* in line 18 most nearly means
   a. essentially.
   b. complicated.
   c. wealthy.
   d. snobby.
   e. mysteriously

161. According to lines 12–25, public art is differentiated from private art mainly by
  a. the kind of ideas or emotions it aims to convey to its audience.
  b. its accessibility.
  c. its perceived value.
  d. its importance to the city.
  e. the recognition that artists receive for their work.

162. The use of the word *sequestered* in line 20 suggests that the author feels
  a. private art is better than public art.
  b. private art is too isolated from the public.
  c. the admission fees for public art arenas prevent many people from experiencing the art.
  d. private art is more difficult to understand than public art.
  e. private art is often controversial in nature.

163. According to lines 33–42, public art serves all of the following functions EXCEPT
  a. beautification.
  b. creation of landmarks.
  c. the fostering of community.
  d. the promotion of good citizenship.
  e. education.

164. Which sentence best sums up the main idea of the passage?
  a. Public art serves several important functions in the city.
  b. Public art is often in direct competition with private art.
  c. Public art should be created both by and for members of the community.
  d. In general, public art is more interesting than private art.
  e. Few people are aware of how much public art is around them.

165. The author's goals in this passage include all of the following EXCEPT
  a. to make readers more aware of the public art works.
  b. to explain the difference between public art and private art.
  c. to explain how public art impacts the city.
  d. to inspire readers to become public artists.
  e. to argue that public art should be created by artists from both inside and outside the community.

166. Which of the following does the author NOT provide in this passage?
   a. an explanation of how the city affects art
   b. specific examples of urban art
   c. a reason why outsiders should create public art
   d. a clear distinction between public and private art
   e. an explanation of how public art regenerates the community

167. Given the author's main purpose, which of the following would most strengthen the passage?
   a. a more detailed discussion of the differences between public and private art.
   b. specific examples of art that fulfills each of the functions discussed in paragraph 5 (lines 33–42).
   c. interviews with public artists about how public art should be created.
   d. a specific example of public art created by a community member versus one created by an outsider to expand paragraph 6 (lines 43–52).
   e. a brief lesson in how to interpret art.

Answers

113. c. The description of the winding paths, shifting landscape and sections that *spill into one another* support the assertion that the park lacks a center.

114. e. Line 8 states that Olmsted wanted to create a *democratic playground*, so he designed the park to have many centers that would *allow interaction among the various members of society* (lines 10–11).

115. b. Line 6 states that the park's design was *innovative*, suggesting it was very different from other park designs.

116. a. Olmsted's goal of creating a democratic park with many centers that would allow interaction among everyone *without giving preference to one group or class* (line 11) shows his philosophy of *inclusion*.

117. b. Lines 3–4 state that the goods *pertaining to the soul* are called *goods in the highest and fullest sense*.

118. d. In line 5 Aristotle notes that the definition of good *corresponds* with the current opinion about the nature of the soul.

119. a. In the second paragraph, Aristotle states that *we have all but defined happiness as a kind of good life and well-being*. Thus, the definitions of happiness and goodness are intertwined; living a good life will bring happiness.

120. **c.** In the third paragraph, Aristotle lists several different ways that people define *happiness* to show that they all fit into the broad definition of *a kind of good life and well-being*.

121. **e.** The opening sentence tells readers that making a list of pros and cons is a technique of utilitarian reasoning. Thus, readers who have used this technique will realize they are already familiar with the basic principles of utilitarianism.

122. **b.** The second sentence explains the main argument of utilitarianism—that we should use consequences to determine our course of action. Thus *posits* is used here in the sense of *asserts*.

123. **c.** Lines 2–4 explain that according to utilitarianism, only the *consequences* of our actions are morally relevant. Lines 5–8 explain that an action is morally good if it creates good (happiness).

124. **d.** Lines 15–17 state the utilitarian principle of choosing actions that create *the greatest amount of good (happiness) for the greatest number of people*.

125. **a.** Lines 17–22 explain two aspects of utilitarianism that complicate the decision-making process: that it is not always clear what the consequences of an action will be (whether they will bring short- or long-term happiness and to what degree), and that sometimes we must sacrifice the happiness of others.

126. **b.** In the first sentence, the author states that *the subject-matter of knowledge is intimately united* (line 2), while in the second sentence he adds *the Sciences [. . .] have multiple bearings on one another* (lines 3–4). In line 6 he states that the sciences *complete, correct, balance each other*.

127. **d.** In the first sentence, the author states that *all branches of knowledge are connected together* (line 1). Then, in the second sentence, he writes *Hence it is that the Sciences, into which our knowledge may be said to be cast* (lines 3–4). Thus, Newman is using the term *the Sciences* to refer to *all branches of knowledge*.

128. **c.** The word *excise* here is used in an unusual way to mean *impose* or *put upon*. The main context clue is the word *influence*, which suggests a *giving to* rather than a *taking away*.

129. **a.** Throughout the first paragraph, the author emphasizes the interdependence of the branches of knowledge and warns against focusing on one branch at the neglect of others. He states that to *give undue prominence to one [area of study] is to be unjust to another; to neglect or supersede these is to divert those from their proper object* (lines 10–12). More importantly, he states that this action would serve to *unsettle the boundary lines between science and science, to destroy the harmony which binds them together* (lines 12–14). Thus the knowledge received would be skewed; it would *tel[l] a different tale* when it is not viewed *as a portion of a whole* (lines 16–17).

130. b. The first sentence of the second paragraph shows that its purpose is to further develop the idea in the first by way of example. Newman writes, *Let me make use of an illustration* (line 19)—an illustration that further demonstrates how one's understanding of an idea changes in relation to the other ideas around it.

131. a. Here *apprehends* is used to mean *understands*. In this paragraph, the author describes what it is the university student would learn from his or her professors.

132. c. Throughout the passage, Newman argues that the branches of knowledge are interrelated and should be studied in combination and in relation to each other. He argues against focusing on one science or discipline, and he states that the university student *apprehends the great outlines of knowledge* (line 50), suggesting that he understands the broad issues in many subject areas.

133. b. At the beginning of the third paragraph, Newman states that *it is a great point then to enlarge the range of studies which a University professes* (lines 35–36) and that students would be best served *by living among those and under those who represent the whole circle* (lines 38–39) of knowledge. He argues that students will learn from the atmosphere created by their professors who *adjust together the claims and relations of their respective subjects* and who *learn to respect, to consult, to aid each other* (lines 43–45).

134. b. The passage defines *panopticon* in lines 7–8: *a place in which everything is in full view of others*. The second paragraph repeats this definition in lines 13–14: *every prisoner's cell would be in full view of the guards.*

135. a. In the third paragraph, the author states that *people behave differently when they know they are being watched* (lines 20–21)—and that when we are being watched, or even think we are being watched, *we will act the way we think we should act when we are being observed by others* (lines 24–25). Thus, the panopticon would be a useful tool for social control. If prisoners know they may be being watched by guards, it is logical to conclude that they are less likely to commit any wrongdoings; thus, the panopticon helps maintain order.

136. c. The author states in line 27 that *the panopticon is already here* and then states that *surveillance cameras are everywhere and we often don't even know our actions are being recorded* (lines 27–29). The rest of the paragraph provides additional examples of how our *cyber-whereabouts* are observed and recorded.

137. d. In Bentham's panopticon, the prisoners *would know they were being watched—or rather, they would know that they could be being watched* (lines 15–17). However, in our modern panopticon, the author states, *we often don't even know our actions are being recorded* (lines 28–29).

138. a. Although information from our credit card purchases is often recorded and exchanged, the author makes no mention of an increased use of credit card purchases contributing to the erosion of privacy. All of the other options, however, are listed in the fourth and sixth paragraphs.

139. c. The paragraph describing the author's experience with identity theft immediately follows the sentence: *We can do little to stop the information gathering and exchange and can only hope to be able to control the damage if something goes wrong* (lines 43–45) and serves as an example of something *going wrong*—the misuse of private information.

140. e. The example of identity theft makes it clear that in cyberspace, *with so much information floating about [. . .] and so much technology that can record and observe* (lines 53–55), our privacy is in jeopardy—it is constantly at risk of being exploited.

141. d. Because of the author's personal experience with identity theft, and because the author finds it *truly amazing that someone would want to live in a transparent house* (lines 56–57), it can be inferred that the author greatly values privacy. The passage also expresses great concern for the lack of control over information in cyberspace (paragraph 4), stating that we *can only hope to be able to control the damage if something goes wrong* (line 44–45). Thus the author would likely support stricter regulations for information gathering and exchange, especially on the Internet.

142. e. In the second sentence the author states that Prometheus is a *complex character*, and in this and the following sentence, the author lists several specific examples of the *rich combination of often-contradictory characteristics* of Prometheus.

143. d. The passage relates the key episodes in the life of Prometheus. This is the only idea broad enough and relevant enough to be the main idea of the passage.

144. b. Prometheus's actions show that he cared for humans more than he cared for Zeus. He gave man knowledge of the arts and sciences although Zeus wanted men to be kept in ignorance (lines 17–18); he tricked Zeus to give mankind the best meat from an ox (line 22); and he stole fire from Mt. Olympus to give mortals the fire that Zeus had denied them (lines 30–31).

145. a. Zeus had given Prometheus and his brother the task of creating humans as a reward for their help in defeating the Titans (lines 7–10).

146. a. Prometheus helped create mortals and then became their *benefactor and protector* (line 15). He is thus most like a *parent* to humans.

147. e. The *transgression* refers back to the previous paragraph, which describes how Prometheus disobeyed Zeus and stole fire from Mount Olympus to give it to man.

148. b. The inclusion of Hope in the jar suggests that Zeus had some pity on mankind and that he wanted to send something to help humans battle the numerous evils he unleashed upon them.

149. c. The style is neither formal nor informal but an easy-going in between to make the material easily understood and interesting to a lay audience. In addition, the passage does not take for granted that the reader knows basic information about mythology. For example, line 9 states that Zeus was the *great ruler of Olympian gods*.

150. d. The members of the PRB were young artists who suddenly found themselves leading a rebellion that had a *dramatic influence on the art world for generations to come* (lines 12–13). The concluding paragraph repeats this idea, stating that these three young men *had a tremendous influence on an entire generation of artists* (lines 58–59). Because *upstart* precedes *young*, we can infer that these men, like the leaders of other rebellions, were suddenly thrown into the spotlight, raised to a high (albeit controversial) position in the art world.

151. d. The author cites the PRB as an example of a rebellion led by *young activists* (line 5) and states that the PRB had a *dramatic influence on the art world* because of their *disdain for the artistic conventions of the time* (line 12). This suggests that their ideas about art were revolutionary, creating a significant and lasting change in the art world. That they were passionate about their beliefs is clear from the fact that they felt strongly enough to form an association and lead a rebellion.

152. **b.** Line 11 states that the oldest PRB member was only 21 years old, so it is clear that the members were young and still developing their skills as artists.

153. **e.** In the third paragraph (lines 14–26), the author states that the PRB believed their peers' art *lack[ed] in meaning and aesthetic honesty* because it often depicted *overly idealized landscapes, carefully arranged family portraits and still lifes, and overly dramatic nature scenes.* In contrast, the PRB believed art should *more accurately depic[t] reality* and portray people, places, and things realistically instead of in an idealized way.

154. **a.** Lines 34–36 state that the PRB's *realism—especially as it related to the Biblical figures—was not well received by many in the art world at the time*.

155. **c.** Lines 14–16 state that the PRB *was formed in response to the brotherhood's belief that the current popular art being produced in England was lacking in meaning and aesthetic honesty*. In addition, line 24 states that the PRB was *committed to bringing greater integrity to art*, suggesting that their peers' work did not have integrity.

156. **e.** The topic sentence of the sixth paragraph states that *one of the most distinctive aspects of PRB works—both in contrast to the works produced during the early nineteenth century and with the art of today—is their dramatic use of color* (lines 45–47).

157. **b.** Throughout the passage, the author describes the principles of the PRB—why the group was formed (paragraphs 2 and 3) and how the group attempted to live up to its principles (paragraphs 4–6). There is little or no information offered about the other answer choices.

158. **a.** In the third paragraph, the author states that the PRB rejected the style and subjects of the Royal Academy, seeking instead *subjects that, by their very nature, had greater meaning and more accurately depicted reality* (lines 22–23). In paragraph four, the author describes how the PRB chose its subjects and aimed to portray people more realistically, thus implying that the members of the PRB had a greater awareness of social issues. In addition, in lines 38–39, the author states that the PRB often chose subjects that *highlight[ed] the societal and moral challenges of the time*.

159. **e.** The three examples in the first paragraph show that there is a wide range of styles of public art in New York City and that public art can be found in a variety of places, including more mundane locations such as the subway and post office.

160. **a.** *Inherently* is an adverb that describes the essential nature of something. The context clue to answer this question is found in the same sentence. *All art is inherently public* because it is *created in order to convey an idea or emotion to others.* The author is saying that an *essential* characteristic of art is that it is created for others.

161. **b.** Line 16 defines public art as *the kind of art created for and displayed in public spaces*, and lines 20–22 state that public art is *specifically designed for a public arena where the art will be encountered by people in their normal day-to-day activities.* This is in contrast to private art, which is less accessible because it is kept in specific, non-public places such as museums and galleries.

162. **b.** To *sequester* is to seclude or isolate. Thus, the use of this word suggests that the author feels private art is too isolated, and cut off from the public.

163. **d.** The seven functions are listed in the fifth paragraph: educating, place making, stimulating the public, promoting community, beautifying, and regenerating. While promoting good citizenship may be a side benefit of public art, it is not discussed in the passage.

164. **a.** After defining public art, the rest of the passage discusses the functions of public art and its impact on the city.

165. **d.** The examples in the first paragraph and the list of different kinds of public art (e.g., ornamental benches in line 28) will make readers more aware of public art; paragraphs 2 and 3 explain the difference between public and private art; paragraph 5 explains how public art affects the community; and paragraph 6 discusses how public art should be created. A few readers may be inspired to create public art after reading this passage, but that is not one of its goals.

166. **a.** Although lines 12–13 states that *there exists in every city a symbiotic relationship between the city and its art* and paragraph 5 explains how public art affects the city, there is no discussion of how the city affects art.

167. **b.** Because the main purpose is to show what public art is and how public art affects the city, the passage would be best served by an expanded discussion of how public art fulfills each of the important functions in paragraph 5.

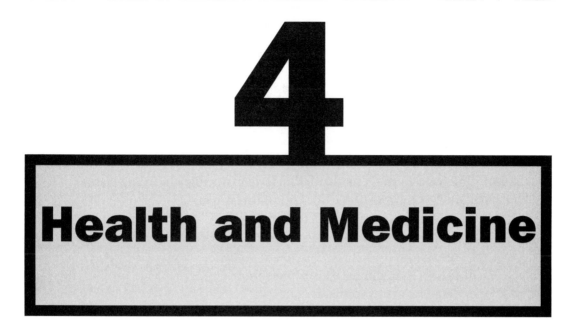

Health and Medicine

Questions 168–171 are based on the following passage.

The following passage is an excerpt from the National Institutes of Health that describes the effects and potential consequences of sleep deprivation.

(1)   Experts say that if you feel drowsy during the day, even during boring activities, you haven't had enough sleep. If you routinely fall asleep within five minutes of lying down, you probably have severe sleep deprivation, possibly even a sleep disorder. *Microsleeps*, or very brief
(5)   episodes of sleep in an otherwise awake person, are another mark of sleep deprivation. In many cases, people are not aware that they are experiencing microsleeps. The widespread practice of "burning the candle at both ends" in Western industrialized societies has created so much sleep deprivation that what is really abnormal sleepiness is now
(10) almost the norm.

    Many studies make it clear that sleep deprivation is dangerous. Sleep-deprived people who are tested by using a driving simulator or by performing a hand-eye coordination task perform as badly as or worse than those who are intoxicated. Sleep deprivation also magni-
(15) fies alcohol's effects on the body, so a fatigued person who drinks will become much more impaired than someone who is well rested. Driver fatigue is responsible for an estimated 100,000 motor vehicle accidents and 1,500 deaths each year, according to the National Highway Traf-

fic Safety Administration. Since drowsiness is the brain's last step
(20) before falling asleep, driving while drowsy can—and often does—lead
to disaster. Caffeine and other stimulants cannot overcome the effects
of severe sleep deprivation. The National Sleep Foundation says that
if you have trouble keeping your eyes focused, if you can't stop yawn-
ing, or if you can't remember driving the last few miles, you are prob-
(25) ably too drowsy to drive safely.

168. The passage suggests that falling asleep during a morning class
   a. means that the topic does not interest you.
   b. is a symptom of sleep deprivation.
   c. indicates that you should drink a caffeinated beverage at
      breakfast.
   d. means that you have a sleep disorder.
   e. requires a visit to the doctor.

169. The image of *burning the candle at both ends* (lines 7–8) most nearly
   refers to
   a. an unrelenting schedule that affords little rest.
   b. an ardent desire to achieve.
   c. the unavoidable conflagration that occurs when two forces
      oppose each other.
   d. a latent period before a conflict or collapse.
   e. a state of extreme agitation.

170. In line 16, the term *impaired* most nearly means
   a. sentient.
   b. apprehensive.
   c. disturbed.
   d. blemished.
   e. hampered.

171. The primary purpose of the passage is to
   a. offer preventive measures for sleep deprivation.
   b. explain why sleeplessness has become a common state in West-
      ern cultures.
   c. recommend the amount of sleep individuals need at different
      ages.
   d. alert readers to the signs and risks of not getting enough sleep.
   e. discuss the effects of alcohol on a sleep-deprived person.

Questions 172–175 refer to the following passage.

In the following passage, the author gives an account of the scientific discoveries made by Antoni van Leeuwenhoek in the fifteenth century.

(1)   The history of microbiology begins with a Dutch haberdasher named Antoni van Leeuwenhoek, a man of no formal scientific education. In the late 1600s, Leeuwenhoek, inspired by the magnifying lenses used by drapers to examine cloth, assembled some of the first microscopes.

(5)   He developed a technique for grinding and polishing tiny, convex lenses, some of which could magnify an object up to 270 times. After scraping some plaque from between his teeth and examining it under a lens, Leeuwenhoek found tiny squirming creatures, which he called "animalcules." His observations, which he reported to the Royal Soci-

(10) ety of London, are among the first descriptions of living bacteria. Leeuwenhoek discovered an entire universe invisible to the naked eye. He found more animalcules—protozoa and bacteria—in samples of pond water, rain water, and human saliva. He gave the first description of red corpuscles, observed plant tissue, examined muscle, and inves-

(15) tigated the life cycle of insects.

    Nearly two hundred years later, Leeuwenhoek's discovery of microbes aided French chemist and biologist Louis Pasteur to develop his "germ theory of disease." This concept suggested that disease derives from tiny organisms attacking and weakening the body. The germ the-

(20) ory later helped doctors to fight infectious diseases including anthrax, diphtheria, polio, smallpox, tetanus, and typhoid. Leeuwenhoek did not foresee this legacy. In a 1716 letter, he described his contribution to science this way: "My work, which I've done for a long time, was not pursued in order to gain the praise I now enjoy, but chiefly from a craving

(25) after knowledge, which I notice resides in me more than in most other men. And therewithal, whenever I found out anything remarkable, I have thought it my duty to put down my discovery on paper, so that all ingenious people might be informed thereof."

172. According to the passage, Leeuwenhoek would be best described as a
    a. bored haberdasher who stumbled upon scientific discovery.
    b. trained researcher with an interest in microbiology.
    c. proficient hobbyist who made microscopic lenses for entertainment.
    d. inquisitive amateur who made pioneer studies of microbes.
    e. talented scientist interested in finding a cure for disease.

173. In line 3, *inspired* most nearly means
  a. introduced.
  b. invested.
  c. influenced.
  d. indulged.
  e. inclined.

174. The quotation from Leeuwenhoek (lines 23–28) is used to illustrate
  a. the value he placed on sharing knowledge among scientists.
  b. that scientific discoveries often go unrecognized.
  c. that much important research is spurred by professional ambition.
  d. the serendipity of scientific progress.
  e. the importance of Leeuwenhoek's discoveries in fighting infectious diseases.

175. The author's attitude toward Leeuwenhoek's contribution to medicine is one of
  a. ecstatic reverence.
  b. genuine admiration.
  c. tepid approval.
  d. courteous opposition.
  e. antagonistic incredulity.

Questions 176–179 are based on the following passage.

The following passage discusses the findings of several recent health surveys investigating the physical activity level of American adolescents.

(1)   Almost 50% of American teens are not vigorously active on a regular basis, contributing to a trend of sluggishness among Americans of all ages, according the U.S. Centers for Disease Control (CDC). Adolescent female students are particularly inactive—29% are inactive
(5)   compared with 15% of male students. Unfortunately, the sedentary habits of young "couch potatoes" often continue into adulthood. According to the Surgeon General's 1996 Report on Physical Activity and Health, Americans become increasingly less active with each year of age. Inactivity can be a serious health risk factor, setting the
(10)   stage for obesity and associated chronic illnesses like heart disease or diabetes. The benefits of exercise include building bone, muscle, and

joints, controlling weight, and preventing the development of high blood pressure.

(15)   Some studies suggest that physical activity may have other benefits as well. One CDC study found that high school students who take part in team sports or are physically active outside of school are less likely to engage in risky behaviors, like using drugs or smoking. Physical activity does not need to be strenuous to be beneficial. The CDC recommends moderate, daily physical activity for people of all ages, such (20) as brisk walking for 30 minutes or 15–20 minutes of more intense exercise. A survey conducted by the National Association for Sport and Physical Education questioned teens about their attitudes toward exercise and about what it would take to get them moving. Teens chose friends (56%) as their most likely motivators for becoming more (25) active, followed by parents (18%) and professional athletes (11%).

176. The first paragraph (lines 1–13) of the passage serves all of the following purposes EXCEPT to
   a. provide statistical information to support the claim that teenagers do not exercise enough.
   b. list long-term health risks associated with lack of exercise.
   c. express skepticism that teenagers can change their exercise habits.
   d. show a correlation between inactive teenagers and inactive adults.
   e. highlight some health benefits of exercise.

177. In line 5, *sedentary* most nearly means
   a. slothful.
   b. apathetic.
   c. stationary.
   d. stabilized.
   e. inflexible.

178. Which of the following techniques is used in the last sentence of the passage (lines 23–25)?
   a. explanation of terms
   b. comparison of different arguments
   c. contrast of opposing views
   d. generalized statement
   e. illustration by example

179. The primary purpose of the passage is to
  a. refute an argument.
  b. make a prediction.
  c. praise an outcome.
  d. promote a change.
  e. justify a conclusion.

Questions 180–187 are based on the following passage.

The following passage discusses the inspiration and career of the first woman to receive a M.D. degree from an American medical school in the nineteenth century.

(1)   Elizabeth Blackwell was the first woman to receive an M.D. degree since the Renaissance, graduating from Geneva Medical College, in New York state, in 1849. She supported women's medical education and helped many other women's careers. By establishing the New York

(5)   Infirmary in 1857, she offered a practical solution to one of the problems facing women who were rejected from internships elsewhere but determined to expand their skills as physicians. She also published several important books on the issue of women in medicine, including *Address on the Medical Education of Women* in 1864 and *Medicine as a*

(10) *Profession for Women* in 1860.

   Elizabeth Blackwell was born in Bristol, England in 1821, to Hannah Lane and Samuel Blackwell. Both for financial reasons and because her father wanted to help abolish slavery, the family moved to America when Elizabeth was eleven years old. Her father died in 1838.

(15) As adults, his children campaigned for women's rights and supported the anti-slavery movement. In her book *Pioneer Work in Opening the Medical Profession to Women*, published in 1895, Dr. Blackwell wrote that she was initially repelled by the idea of studying medicine. She said she had "hated everything connected with the body, and could not

(20) bear the sight of a medical book . . . My favorite studies were history and metaphysics, and the very thought of dwelling on the physical structure of the body and its various ailments filled me with disgust." Instead she went into teaching, then considered more suitable for a woman. She claimed that she turned to medicine after a close friend

(25) who was dying suggested she would have been spared her worst suffering if her physician had been a woman.

   Blackwell had no idea how to become a physician, so she consulted with several physicians known by her family. They told her it was a fine idea, but impossible; it was too expensive, and such education was

(30)   not available to women. Yet Blackwell reasoned that if the idea were
a good one, there must be some way to do it, and she was attracted by
the challenge. She convinced two physician friends to let her read
medicine with them for a year, and applied to all the medical schools
in New York and Philadelphia. She also applied to twelve more
(35)   schools in the northeast states and was accepted by Geneva Medical
College in 1847. The faculty, assuming that the all-male student body
would never agree to a woman joining their ranks, allowed them to
vote on her admission. As a joke, they voted "yes," and she gained
admittance, despite the reluctance of most students and faculty.

(40)   Two years later, in 1849, Elizabeth Blackwell became the first
woman to receive an M.D. degree from an American medical school.
She worked in clinics in London and Paris for two years, and studied
midwifery at La Maternité where she contracted "purulent opthalmia"
from a young patient. When Blackwell lost sight in one eye, she
(45)   returned to New York City in 1851, giving up her dream of becoming a surgeon.

       Dr. Elizabeth Blackwell established a practice in New York City, but
had few patients and few opportunities for intellectual exchange with
other physicians and "the means of increasing medical knowledge
(50)   which dispensary practice affords." She applied for a job as physician
at the women's department of a large city dispensary, but was refused.
In 1853, with the help of friends, she opened her own dispensary in a
single rented room, seeing patients three afternoons a week. The dispensary was incorporated in 1854 and moved to a small house she
(55)   bought on 15th Street. Her sister, Dr. Emily Blackwell, joined her in
1856 and, together with Dr. Marie Zakrzewska, they opened the New
York Infirmary for Women and Children at 64 Bleecker Street in
1857. This institution and its medical college for women (opened
1867) provided training and experience for women doctors and med-
(60)   ical care for the poor.

       As her health declined, Blackwell gave up the practice of medicine
in the late 1870s, though she still campaigned for reform.

180.   The passage is primarily concerned with
   a. the inevitable breaking down of social barriers for women.
   b. the effect of adversity in shaping a person's life.
   c. one woman's determination to open the field of medicine to
        females.
   d. one woman's desire to gain prestige.
   e. the quality of healthcare available in the 1800s.

181. The word *practical* (line 5) most nearly means
   a. usable.
   b. satisfactory.
   c. systematic.
   d. professional.
   e. adept.

182. The author mentions Samuel Blackwell's involvement in the anti-slavery movement (lines 13–14) in order to
   a. offer random biographical information about Elizabeth's upbringing.
   b. suggest that her father's beliefs greatly influenced Elizabeth.
   c. imply a link between financial need and the abhorrence of slavery.
   d. describe the political ferment that preceded the American Civil War.
   e. explain Elizabeth's choice of medicine for a profession.

183. In line 18, the word *repelled* most nearly means
   a. vanquished.
   b. discouraged.
   c. intimidated.
   d. depressed.
   e. sickened.

184. According to the passage, Blackwell chose to become a doctor
   a. as a result of the encouragement of physicians known to her family.
   b. despite the fact that most considered her goal inappropriate and unattainable.
   c. in order to make healthcare more accessible to the poor.
   d. because she hoped to overcome her revulsion of the body and disease.
   e. to fulfill a childhood dream of establishing a medical college for women.

185. As described in lines 36–39, the actions of the student body of Geneva Medical College suggest that they
   a. admired Blackwell's ambition.
   b. respected the politics of the Blackwell family.
   c. doubted Blackwell's commitment to medicine.
   d. feared the influence of Blackwell's family connections.
   e. made light of Blackwell's goal.

186. The passage implies that Blackwell's attitude toward studying and practicing medicine changed from
  a. tenacious to wavering.
  b. uninterested to resolute.
  c. cynical to committed.
  d. idealized to realistic.
  e. theoretical to practical.

187. All of the following questions can be explicitly answered on the basis of the passage EXCEPT
  a. What barriers did Blackwell face in her pursuit to become a physician?
  b. What degree of success did women attain in the field of medicine as a result of Blackwell?
  c. What contributions did Blackwell make to women interested in medicine as a profession?
  d. What specific steps did Blackwell take to gain admittance to medical school?
  e. What did Blackwell claim was her inspiration for wanting to become a doctor?

Questions 188–195 are based on the following passage.

The following passage offers the author's perspective on the need for healthcare providers with specialized training to care for a rapidly expanding population of older Americans.

(1)   The U.S. population is going gray. A rising demographic tide of aging baby boomers—those born between 1946 and 1964—and increased longevity have made adults age 65 and older the fastest growing segment of today's population. In thirty years, this segment of the popu-

(5)   lation will be nearly twice as large as it is today. By then, an estimated 70 million people will be over age 65. The number of "oldest old"— those age 85 and older—is 34 times greater than in 1900 and likely to expand five-fold by 2050.

         This unprecedented "elder boom" will have a profound effect on

(10)   American society, particularly the field of healthcare. Is the U.S. health system equipped to deal with the demands of an aging population? Although we have adequate physicians and nurses, many of them are not trained to handle the multiple needs of older patients. Today we have about 9,000 geriatricians (physicians who are experts in aging-

(15)   related issues). Some studies estimate a need for 36,000 geriatricians by 2030.

Many doctors today treat a patient of 75 the same way they would treat a 40–year-old patient. However, although seniors are healthier than ever, physical challenges often increase with age. By age 75, *(20)* adults often have two to three medical conditions. Diagnosing multiple health problems and knowing how they interact is crucial for effectively treating older patients. Healthcare professionals—often pressed for time in hectic daily practices—must be diligent about asking questions and collecting "evidence" from their elderly patients. Finding *(25)* out about a patient's over-the-counter medications or living conditions could reveal an underlying problem.

Lack of training in geriatric issues can result in healthcare providers overlooking illnesses or conditions that may lead to illness. Inadequate nutrition is a common, but often unrecognized, problem among frail *(30)* seniors. An elderly patient who has difficulty preparing meals at home may become vulnerable to malnutrition or another medical condition. Healthcare providers with training in aging issues may be able to address this problem without the costly solution of admitting a patient to a nursing home.

(35)   Depression, a treatable condition that affects nearly five million seniors, also goes undetected by some healthcare providers. Some healthcare professionals view depression as "just part of getting old." Untreated, this illness can have serious, even fatal consequences. According to the National Institute of Mental Health, older Ameri- *(40)* cans account for a disproportionate share of suicide deaths, making up 18% of suicide deaths in 2000. Healthcare providers could play a vital role in preventing this outcome—several studies have shown that up to 75% of seniors who die by suicide visited a primary care physician within a month of their death.

(45)   Healthcare providers face additional challenges to providing high-quality care to the aging population. Because the numbers of ethnic minority elders are growing faster than the aging population as a whole, providers must train to care for a more racially and ethnically diverse population of elderly. Respect and understanding of diverse *(50)* cultural beliefs is necessary to provide the most effective healthcare to all patients. Providers must also be able to communicate complicated medical conditions or treatments to older patients who may have a visual, hearing, or cognitive impairment.

As older adults make up an increasing proportion of the healthcare *(55)* caseload, the demand for aging specialists must expand as well.

Healthcare providers who work with the elderly must understand and address not only the physical but mental, emotional, and social changes of the aging process. They need to be able to distinguish between "normal" characteristics associated with aging and illness. *(60)* Most crucially, they should look beyond symptoms and consider ways that will help a senior maintain and improve her quality of life.

188. The author uses the phrase *going gray* (line 1) in order to
   a. maintain that everyone's hair loses its color eventually.
   b. suggest the social phenomenon of an aging population.
   c. depict older Americans in a positive light.
   d. demonstrate the normal changes of aging.
   e. highlight the tendency of American culture to emphasize youth.

189. The tone of the passage is primarily one of
   a. bemused inquiry.
   b. detached reporting.
   c. informed argument.
   d. hysterical plea.
   e. playful speculation.

190. The author implies that doctors who treat an elderly patient the same as they would *a 40–year-old patient* (line 18)
   a. provide equitable, high-quality care.
   b. avoid detrimental stereotypes about older patients.
   c. encourage middle-age adults to think about the long-term effects of their habits.
   d. do not offer the most effective care to their older patients.
   e. willfully ignore the needs of the elderly.

191. In line 33, the word *address* most nearly means
   a. manage.
   b. identify.
   c. neutralize.
   d. analyze.
   e. dissect.

192. The author cites the example of untreated depression in elderly people (lines 35–38) in order to
  a. prove that mental illness can affect people of all ages.
  b. undermine the perception that mental illness only affects young people.
  c. support the claim that healthcare providers need age-related training.
  d. show how mental illness is a natural consequence of growing old.
  e. illustrate how unrecognized illnesses increase the cost of healthcare.

193. According to the passage, which of the following is NOT a possible benefit of geriatric training for healthcare providers?
  a. improved ability to explain a medical treatment to a person with a cognitive problem
  b. knowledge of how heart disease and diabetes may act upon each other in an elderly patient
  c. improved ability to attribute disease symptoms to the natural changes of aging
  d. more consideration for ways to improve the quality of life for seniors
  e. increased recognition of and treatment for depression in elders

194. The author implies that a healthcare system that routinely looks *beyond symptoms* (line 60) is one that
  a. intrudes on the private lives of individuals.
  b. considers more than just the physical aspects of a person.
  c. rivals the social welfare system.
  d. misdiagnoses diseases that are common in the elderly.
  e. promotes the use of cutting-edge technology in medical care.

195. In the last paragraph of the passage (lines 54–61) the author's tone is one of
  a. unmitigated pessimism.
  b. personal reticence.
  c. hypocritical indifference.
  d. urgent recommendation.
  c. frenzied panic.

Questions 196–203 are based on the following passage.

The following passage is an excerpt from a recent introduction to the momentous 1964 Report on Smoking and Health issued by the United States Surgeon General. It discusses the inspiration behind the report and the report's effect on public attitudes toward smoking.

(1)   No single issue has preoccupied the Surgeons General of the past four decades more than smoking. The reports of the Surgeon General have alerted the nation to the health risk of smoking, and have transformed the issue from one of individual and consumer choice, to one of epi-
(5)   demiology, public health, and risk for smokers and non-smokers alike.

        Debate over the hazards and benefits of smoking has divided physicians, scientists, governments, smokers, and non-smokers since *Tobacco nicotiana* was first imported to Europe from its native soil in the Americas in the sixteenth century. A dramatic increase in cigarette
(10)   smoking in the United States in the twentieth century called forth anti-smoking movements. Reformers, hygienists, and public health officials argued that smoking brought about general malaise, physiological malfunction, and a decline in mental and physical efficiency. Evidence of the ill effects of smoking accumulated during the 1930s,
(15)   1940s, and 1950s.

        Epidemiologists used statistics and large-scale, long-term, case-control surveys to link the increase in lung cancer mortality to smoking. Pathologists and laboratory scientists confirmed the statistical relationship of smoking to lung cancer as well as to other serious dis-
(20)   eases, such as bronchitis, emphysema, and coronary heart disease. Smoking, these studies suggested, and not air pollution, asbestos contamination, or radioactive materials, was the chief cause of the epidemic rise of lung cancer in the twentieth century. On June 12, 1957, Surgeon General Leroy E. Burney declared it the official position of
(25)   the U.S. Public Health Service that the evidence pointed to a causal relationship between smoking and lung cancer.

        The impulse for an official report on smoking and health, however, came from an alliance of prominent private health organizations. In June 1961, the American Cancer Society, the American Heart Asso-
(30)   ciation, the National Tuberculosis Association, and the American Public Health Association addressed a letter to President John F. Kennedy, in which they called for a national commission on smoking, dedicated to "seeking a solution to this health problem that would interfere least with the freedom of industry or the happiness of individuals." The
(35)   Kennedy administration responded the following year, after prompting from a widely circulated critical study on cigarette smoking by the

Royal College of Physicians of London. On June 7, 1962, recently
appointed Surgeon General Luther L. Terry announced that he would
convene a committee of experts to conduct a comprehensive review of
(40)   the scientific literature on the smoking question. . . .

Meeting at the National Library of Medicine on the campus of the
National Institutes of Health in Bethesda, Maryland, from November
1962 through January 1964, the committee reviewed more than 7,000
scientific articles with the help of over 150 consultants. Terry issued
(45)   the commission's report on January 11, 1964, choosing a Saturday to
minimize the effect on the stock market and to maximize coverage in
the Sunday papers. As Terry remembered the event, two decades later,
the report "hit the country like a bombshell. It was front page news
and a lead story on every radio and television station in the United
(50)   States and many abroad."

The report highlighted the deleterious health consequences of
tobacco use. *Smoking and Health: Report of the Advisory Committee to the
Surgeon General* held cigarette smoking responsible for a 70% increase
in the mortality rate of smokers over non-smokers. The report esti-
(55)   mated that average smokers had a nine- to ten-fold risk of developing
lung cancer compared to non-smokers: heavy smokers had at least a
twenty-fold risk. The risk rose with the duration of smoking and
diminished with the cessation of smoking. The report also named
smoking as the most important cause of chronic bronchitis and
(60)   pointed to a correlation between smoking and emphysema, and smoking and coronary heart disease. It noted that smoking during pregnancy reduced the average weight of newborns. On one issue the
committee hedged: nicotine addiction. It insisted that the "tobacco
habit should be characterized as an habituation rather than an addic-
(65)   tion," in part because the addictive properties of nicotine were not yet
fully understood, in part because of differences over the meaning of
addiction.

The 1964 report on smoking and health had an impact on public
attitudes and policy. A Gallup Survey conducted in 1958 found that
(70)   only 44% of Americans believed smoking caused cancer, while 78%
believed so by 1968. In the course of a decade, it had become common
knowledge that smoking damaged health, and mounting evidence of
health risks gave Terry's 1964 report public resonance. Yet, while the
report proclaimed that "cigarette smoking is a health hazard of suffi-
(75)   cient importance in the United States to warrant appropriate remedial
action," it remained silent on concrete remedies. That challenge fell
to politicians. In 1965, Congress required all cigarette packages dis-

tributed in the United States to carry a health warning, and since 1970
this warning is made in the name of the Surgeon General. In 1969,
(80) cigarette advertising on television and radio was banned, effective September 1970.

196. The primary purpose of the passage is to
   a. show the mounting evidence of the deleterious health conse-
      quences of smoking.
   b. explain why the Kennedy administration called for a national
      commission on smoking.
   c. describe the government's role in protecting public health.
   d. show the significance of the 1964 Surgeon General's report.
   e. account for the emergence of anti-smoking movements in
      twentieth-century United States.

197. In line 1, *preoccupied* most nearly means
   a. distressed.
   b. beset.
   c. absorbed.
   d. inconvenienced.
   e. fomented.

198. The first sentence of the second paragraph (lines 6–9) is intended
   to express the
   a. long-standing controversy about the effects of smoking.
   b. current consensus of the medical community regarding
      smoking.
   c. government's interest in improving public health.
   d. ongoing colloquy between physicians, scientists, and
      governments.
   e. causal relationship between smoking and lung disease.

199. The author implies that *the impulse* (line 27) to create a
   government report on smoking
   a. was an overdue response to public demand.
   b. would not have been pursued if John F. Kennedy was not
      president.
   c. came from within the U.S. Public Health Service.
   d. would meet with significant opposition from smokers around
      the country.
   e. was the result of pressure from forces outside of the government.

200. The quotation by Surgeon General Luther L. Terry (lines 48–50)
is used to illustrate the
a. outrage of consumers wanting to protect their right to smoke.
b. disproportionate media coverage of the smoking report.
c. overreaction of a hysterical public.
d. explosive response to the revelation of smoking's damaging
effects.
e. positive role government can play in people's lives.

201. In line 63, *hedged* most nearly means
a. exaggerated.
b. evaded.
c. deceived.
d. speculated.
e. hindered.

202. The statement that the 1964 Surgeon General's report remained
silent on concrete remedies (line 76) implies that it
a. served primarily as a manifesto that declared the views of the
Surgeon General.
b. could have recommended banning cigarette advertising but it
did not.
c. was ignorant of possible remedial actions.
d. maintained its objectivity by abstaining from making policy
recommendations.
e. did not deem it necessary to recommend specific actions that
would confront the health problem of smoking.

203. In the last paragraph of the passage, the attitude of the author
toward the legacy of the 1964 Surgeon General's report is one of
a. unqualified praise.
b. appreciation.
c. wonderment.
d. cynicism.
e. disillusionment.

Questions 204–212 are based on the following passages.

These two passages reflect two different views of the value of cosmetic plastic surgery. Passage 1 is an account by a physician who has practiced internal medicine (general medicine) for more than two decades and who has encountered numerous patients inquiring about cosmetic plastic surgery procedures. Passage 2 is written by a professional woman in her mid-forties who has considered cosmetic plastic surgery for herself.

PASSAGE 1

(1)   Elective and cosmetic plastic surgery is one of the fastest growing segments of healthcare, second only to geriatric care. As the "baby boomers" (those born between 1945 and 1965) reach their half-century mark, more Americans are seeking cosmetic procedures that min-
(5)   imize the visible signs of aging. The demand for self-improvement has increased as the job market has become more competitive and a high divorce rate spurs the search for new personal relationships. Increased discretionary wealth and a wider acceptance of cosmetic techniques have also contributed to the spike in cosmetic surgery.
(10)   In the 1980s, I was just beginning as an internist, working in a private practice. Then in my late twenties, I felt pity for my patients who talked to me about a surgical fix for their wrinkles or other signs of aging. I felt that if they had a developed sense of self-esteem, they would not feel the need to surgically alter their appearance. I also felt
(15)   a certain degree of envy for my cosmetic-surgeon colleagues, some of whom worked across the hall. To my "green" eye, they looked like slick salespeople reaping large financial rewards from others' insecurity and vanity. It was difficult for me to reconcile the fact that patients were willing to fork over thousands of dollars for cosmetic fixes, while
(20)   primary care physicians struggled to keep their practices financially viable.
         Since that time, my attitude has changed. Although cosmetic surgery sometimes produces negative outcomes—the media often highlights surgery "disasters"—for the most part, the health risk for
(25)   cosmetic procedures is low and patient satisfaction is high. Often, people who have been hobbled by poor body image all of their lives, walk away from cosmetic surgery with confidence and the motivation to lead healthier lives. In addition, reconstructive surgery for burn and accident victims or to those disfigured from disease restores self-
(30)   esteem and wellbeing in a way that other therapies cannot. I believe

it is time for members of the medical community to examine the benefits and results of cosmetic surgery without prejudice or jealousy.

PASSAGE 2

(1) Beauty is only skin deep, or so goes the old adage. However, in a culture increasingly fixated on youthfulness and saturated with media images of "ideal"-looking men and women, cosmetic plastic surgery seems like the norm instead of the exception. Nearly 6.6 million

(5) Americans opted for cosmetic surgery in 2002, with women accounting for 85% of cosmetic-surgery patients, according to the American Society of Plastic Surgeons. Once the province of older women, cosmetic surgery is increasingly an option for 35– to 50–year-olds, who made up 45% of cosmetic-surgery patients in 2002.

(10) Coming of age in the 1970s, I grew up believing in the spirit of feminism, a ready warrior for equal rights for women in the home and workplace. I believed that women should be valued for who they are and what they do, and not for how they look. But as I approach my mid-forties, I look in the mirror and wonder about the reflection I see.

(15) Although I adhere to a healthy lifestyle, eat well, exercise regularly, and feel energetic, the reality is that I am beginning to look, well, middle-aged.

Because I am a successful professional, I have the means to afford elective surgery. And like Pandora's Box, once I opened the door to

(20) anti-aging surgical possibilities, it seems almost impossible to close it again. In 2002, more than 1.1 million Americans had Botox injections—a procedure that erases wrinkles by paralyzing facial muscles. I find myself asking: Why not me? Is it time to jump on the bandwagon? In a competitive culture where looks count, is it almost

(25) *impractical* not to?

What stops me? Perhaps it is queasiness about the surgeon's scalpel. Risks accompany any kind of surgery. Perhaps I find the idea of paralyzing my facial muscles somewhat repellent and a betrayal of the emotions I have experienced—the joys and loses of a lifetime—that are

(30) written in those "crow's feet" and "worry lines." Perhaps yet, it is my earlier feminist fervor and idealism—a remnant of my youth that I believe is worth preserving more than wrinkle-free skin.

204. The word *adage* (Passage 2, line 1) most nearly means
    a. addition.
    b. rumor.
    c. saying.
    d. era.
    e. fib.

205. The argument of Passage 1 would be most effectively strengthened by which of the following?
    a. information about making plastic surgery more affordable
    b. anecdotes about incompetent plastic surgeons
    c. facts to support the author's claim that health risks are low for cosmetic procedures
    d. a description of the author's personal experience with patients
    e. a description of the psychological benefits of improved body image

206. In the second paragraph of Passage 1 (lines 10–21), how would the author characterize the motivation of cosmetic plastic surgeons?
    a. altruistic
    b. professional
    c. creative
    d. thrilling
    e. greedy

207. Which audience is the author of Passage 1 most likely addressing?
    a. burn or accident victims
    b. women with poor body image
    c. plastic surgeons
    d. healthcare providers
    e. "baby boomers"

208. In Passage 2, line 2 *saturated* most nearly means
    a. animated.
    b. decorated.
    c. gratified.
    d. permeated.
    e. tainted.

209. The author of Passage 2 implies that feminists of the 1970s held which of the following beliefs?
   a. All women should have the right to safe, affordable cosmetic surgery.
   b. Looks should not be a factor in determining a person's worth.
   c. Cosmetic surgery is a beneficial tool in that it increases a woman's self-esteem.
   d. To be fair, men should be judged by their looks, too.
   e. Women should do whatever is necessary to compete in the job market.

210. Which aspect of the cosmetic plastic surgery trend is emphasized in Passage 1, but not in Passage 2?
   a. professional envy among doctors
   b. nonsurgical techniques like Botox injections
   c. media's role in promoting plastic surgery
   d. surgical risks
   e. cost of procedures

211. The two authors would most likely agree with which statement?
   a. Cosmetic surgery takes away individuality.
   b. Ideals of beauty are not culturally informed.
   c. Plastic surgeons prey off of vulnerable patients.
   d. American society is highly competitive.
   e. The benefits of plastic surgery outweigh the risks.

212. The approaches of the two passages to the topic are the similar in that they both use
   a. first-person experiences.
   b. second-person address to the reader.
   c. references to other sources on the subject.
   d. a summary of types of plastic surgery.
   e. statistics on patient satisfaction.

Questions 213–222 are based on the following passage.

This passage describes the public's growing interest in alternative medicine practices in twenty-first century United States.

(1)   Once people wore garlic around their necks to ward off disease. Today, most Americans would scoff at the idea of wearing a necklace of garlic cloves to enhance their wellbeing. However, you might find a number

(5) of Americans willing to ingest capsules of pulverized garlic or other herbal supplements in the name of health.

Complementary and alternative medicine (CAM), which includes a range of practices outside of conventional medicine such as herbs, homeopathy, massage, yoga, and acupuncture, holds increasing appeal for Americans. In fact, according to one estimate, 42% of (10) Americans have used alternative therapies. A Harvard Medical School survey found that young adults (those born between 1965 and 1979) are the most likely to use alternative treatments, whereas people born before 1945 are the least likely to use these therapies. Nonetheless, in all age groups, the use of unconventional healthcare practices has (15) steadily increased since the 1950s, and the trend is likely to continue.

CAM has become a big business as Americans dip into their wallets to pay for alternative treatments. A 1997 American Medical Association study estimated that the public spent $21.2 billion for alternative medicine therapies in that year, more than half of which were "out-of-(20) pocket" expenditures, meaning they were not covered by health insurance. Indeed, Americans made more out-of-pocket expenditures for alternative services than they did for out-of-pocket payments for hospital stays in 1997. In addition, the number of total visits to alternative medicine providers (about 629 million) exceeded the tally of visits (25) to primary care physicians (386 million) in that year.

However, the public has not abandoned conventional medicine for alternative healthcare. Most Americans seek out alternative therapies as a complement to their conventional healthcare whereas only a small percentage of Americans rely primarily on alternative care. Why have (30) so many patients turned to alternative therapies? Frustrated by the time constraints of managed care and alienated by conventional medicine's focus on technology, some feel that a holistic approach to healthcare better reflects their beliefs and values. Others seek therapies that will relieve symptoms associated with chronic disease, symp-(35) toms that mainstream medicine cannot treat.

Some alternative therapies have crossed the line into mainstream medicine as scientific investigation has confirmed their safety and efficacy. For example, today physicians may prescribe acupuncture for pain management or to control the nausea associated with chemother-(40) apy. Most U.S. medical schools teach courses in alternative therapies and many health insurance companies offer some alternative medicine benefits. Yet, despite their gaining acceptance, the majority of alternative therapies have not been researched in controlled studies. New research efforts aim at testing alternative methods and providing the

(45) public with information about which are safe and effective and which are a waste of money, or possibly dangerous.

So what about those who swear by the health benefits of the "smelly rose," garlic?

Observational studies that track disease incidence in different pop-
(50) ulations suggest that garlic use in the diet may act as a cancer-fighting agent, particularly for prostate and stomach cancer. However, these findings have not been confirmed in clinical studies. And yes, reported side effects include garlic odor.

213. The author's primary purpose in the passage is to
  a. confirm the safety and effectiveness of alternative medicine approaches.
  b. convey the excitement of crossing new medical frontiers.
  c. describe the recent increase in the use of alternative therapies.
  d. explore the variety of practices that fall into the category of alternative medicine.
  e. criticize the use of alternative therapies that have not been scientifically tested.

214. The author describes wearing garlic (line 1) as an example of
  a. an arcane practice considered odd and superstitious today.
  b. the ludicrous nature of complementary and alternative medicine.
  c. a scientifically tested medical practice.
  d. a socially unacceptable style of jewelry.
  e. a safe and reliable means to prevent some forms of cancer.

215. The word *conventional* as it is used in line 7 most nearly means
  a. appropriate.
  b. established.
  c. formal.
  d. moralistic.
  e. reactionary.

216. The author most likely uses the Harvard survey results (lines 10–13) to imply that
   a. as people age they always become more conservative.
   b. people born before 1945 view alternative therapies with disdain.
   c. the survey did not question baby boomers (those born between 1945–1965) on the topic.
   d. many young adults are open-minded to alternative therapies.
   e. the use of alternative therapies will decline as those born between 1965 and 1979 age.

217. The statistic comparing total visits to alternative medicine practitioners with those to primary care physicians (lines 23–25) is used to illustrate the
   a. popularity of alternative medicine.
   b. public's distrust of conventional healthcare.
   c. accessibility of alternative medicine.
   d. affordability of alternative therapies.
   e. ineffectiveness of most primary care physicians.

218. In line 28, *complement* most nearly means
   a. tribute.
   b. commendation.
   c. replacement.
   d. substitute.
   e. addition.

219. The information in lines 30–35 indicates that Americans believe that conventional healthcare
   a. offers the best relief from the effects of chronic diseases.
   b. should not use technology in treating illness.
   c. combines caring for the body with caring for the spirit.
   d. falls short of their expectations in some aspects.
   e. needs a complete overhaul to become an effective system.

220. The author suggests that *cross[ing] the line into mainstream medicine* (lines 36–37) involves
   a. performing stringently controlled research on alternative therapies.
   b. accepting the spiritual dimension of preventing and treating illness.
   c. approving of any treatments that a patient is interested in trying.
   d. recognizing the popularity of alternative therapies.
   e. notifying your physician about herbs or alternative therapies you are using.

221. In lines 49–54, the author refers to garlic use again in order to
   a. cite an example of the fraudulent claims of herbal supplements.
   b. suggest that claims about some herbs may be legitimate.
   c. mock people who take garlic capsules.
   d. reason why some Americans are drawn to alternative health methods.
   e. argue that observational studies provide enough evidence.

222. Which of the following best describes the approach of the passage?
   a. matter-of-fact narration
   b. historical analysis
   c. sarcastic criticism
   d. playful reporting
   e. impassioned argument

Questions 223–232 are based on the following passage.

In the following article, the author speculates about a connection between the low-fat, high-carbohydrate diet recommended by the medical establishment in the last twenty years and the increasing rate of obesity among Americans.

(1)   American dietitians and members of the medical community have ridiculed low-carbohydrate diets as quackery for the past thirty years, while extolling a diet that cuts down on fat, limits meat consumption, and relies on carbohydrates as its staple. Many Americans are famil-
(5)   iar with the food pyramid promoted by the U.S. government, with its foundation of carbohydrates such as breads, rice, and pasta, and its apex allotted to fats, oils, and sweets. Adhering to the government's

anti-fat, pro-carbohydrate gospel, food manufacturers have pumped out fat-free grain products that lure consumers with the promise of
(10) leaner days. Then, why are Americans getting so fat? Could the dietary recommendations of the last twenty years be wrong? And what's more, could the proponents of diets that push protein and fat be *right?*

Fact: Obesity rates have soared throughout the country since the
(15) 1980s. The United States Centers of Disease Control reports that the number of obese adults has doubled in the last twenty years. The number of obese children and teenagers has almost tripled, increasing 120% among African-American and Latino children and 50% among white children. The risk for Type 2 diabetes, which is associ-
(20) ated with obesity, has increased dramatically as well. Disturbingly, the disease now affects 25% to 30% of children, compared with 3% to 5% two decades ago.

What is behind this trend? Supersized portions, cheap fast food, and soft drinks combined with a sedentary lifestyle of TV watching or
(25) Internet surfing have most likely

contributed to the rapid rise of obesity. Yet, there might be more to it: is it a coincidence that obesity rates increased in the last twenty years—the same time period in which the low-fat dietary doctrine has reigned? Before the 1980s, the conventional wisdom was that fat and
(30) protein created a feeling of satiation, so that overeating would be less likely. Carbohydrates, on the other hand, were regarded as a recipe for stoutness. This perception began to change after World War II when coronary heart disease reached near epidemic proportions among middle-aged men. A theory that dietary fat might increase cholesterol
(35) levels and, in turn, increase the risk of heart disease emerged in the 1950s and gained increasing acceptance by the late 1970s. In 1979, the focus of the food guidelines promoted by the United States Department of Agriculture (USDA) began to shift away from getting enough nutrients to avoiding excess fat, saturated fat, cholesterol, and
(40) sodium—the components believed to be linked to heart disease. The anti-fat credo was born.

To date, the studies that have tried to link dietary fat to increased risk of coronary heart disease have remained ambiguous. Studies have shown that cholesterol-lowering drugs help reduce the risk of heart
(45) disease, but whether a diet low in cholesterol can do the same is still questionable. While nutrition experts are debating whether a low-fat, carbohydrate-based diet is the healthiest diet for Americans, nearly all agree that the anti-fat message of the last twenty years has been over-simplified. For example, some fats and oils like those found in olive oil

(50) and nuts are beneficial to the heart and may deserve a larger proportion in the American diet than their place at the tip of the food pyramid indicates. Likewise, some carbohydrates that form the basis of the
food pyramid, like the "refined" carbohydrates contained in white
bread, pasta, and white rice, are metabolized in the body much the
(55) same way sweets are. According to one Harvard Medical School
researcher, a breakfast of a bagel with low-fat cream cheese is "meta-
bolically indistinguishable from a bowl of sugar."

So what about those high-fat, protein diets that restrict carbohydrates like the popular Atkins' diet and others? A small group of nutri-
(60) tion experts within the medical establishment find it hard to ignore the
anecdotal evidence that many lose weight successfully on these diets.
They are arguing that those diets should not be dismissed out of hand,
but researched and tested more closely. Still others fear that Americans, hungry to find a weight-loss regimen, may embrace a diet that
(65) has no long-term data about whether it works or is safe. What is clear
is that Americans are awaiting answers and in the meantime, we need
to eat *something*.

223. The passage is primarily concerned with
   a. questioning the dietary advice of the past two decades.
   b. contrasting theories of good nutrition.
   c. displaying the variety of ways one can interpret scientific
      evidence.
   d. debunking the value of diets that restrict carbohydrates.
   e. isolating the cause of the rising rate of obesity.

224. The author's attitude toward the medical experts who *ridiculed low-
carbohydrate diets as quackery* and praised low-fat diets is one of
   a. bemused agreement.
   b. seeming ambivalence.
   c. unconcerned apathy.
   d. implicit objection.
   e. shocked disbelief.

225. The term *gospel* (line 8) as it is used in the passage most nearly
means
   a. one of the first four New Testament books.
   b. a proven principle.
   c. a message accepted as truth.
   d. American evangelical music.
   e. a singular interpretation.

226. The author uses the word *Fact* (line 14) in order to
   a. draw a conclusion about the USDA's dietary recommendations.
   b. imply that statistical information can be misleading.
   c. hypothesize about the health effects of high-fat, protein diets.
   d. introduce a theory about the increased rate of obesity.
   e. emphasize a statistical reality regardless of its cause.

227. The passage suggests that the obesity trend in the United States is
   a. partly a result of inactive lifestyles.
   b. the predictable outcome of cutting down on saturated fat.
   c. a cyclical event that happens every twenty years.
   d. unrelated to a rise in diabetes cases.
   e. the unfortunate byproduct of the effort to reduce heart disease.

228. In lines 26–31, the author implies that the government's 1979 food guidelines
   a. relied more on folk wisdom than on scientific study.
   b. was based on the theoretical premise that eating less dietary fat reduces heart disease.
   c. was negligent in not responding to the increasing incidence of heart disease.
   d. no longer bothered to mention nutrient objectives.
   e. was successful in reducing heart disease rates.

229. The author characterizes the *anti-fat message of the last twenty years* (line 48) as
   a. elusive.
   b. questionable.
   c. incoherent.
   d. beneficial.
   e. inventive.

230. The author cites the example of *a breakfast of a bagel with low-fat cream cheese* in order to
   a. show that getting a nutritional breakfast can be fast and convenient.
   b. demonstrate that carbohydrates are the ideal nutrient.
   c. overturn the notion that a carbohydrate-based breakfast is necessarily healthy.
   d. persuade readers that they should eat eggs and sausage for breakfast.
   e. argue that Americans should greatly restrict their carbohydrate intake.

231. The author of the passage would most likely agree with which statement?
   a. The federal government knowingly gave the public misleading advice.
   b. Soaring obesity rates are most certainly a result of low-fat diets.
   c. Nutritionists should promote high-fat, protein diets like the Atkin's diet.
   d. Scientists should investigate every fad diet with equal scrutiny.
   e. There is no definitive evidence connecting dietary fat to heart disease.

232. The tone of the last sentence of the passage (lines 65–67) is one of
   a. optimism.
   b. resolve.
   c. indulgence.
   d. irony.
   e. revulsion.

Answers

168. **b.** The passage states that daytime drowsiness, *even during boring activities* (lines 1–2), is a sign that a person is not getting enough sleep.

169. **a.** This image connotes a state of working hard without adequate rest.

170. **e.** The passage claims that lack of sleep *magnifies alcohol's effects on the body* (lines 14–15) implying that it hampers a person's ability to function.

171. **d.** The first paragraph of this short passage deals with the symptoms of sleep deprivation and the second paragraph discusses the dangers of not getting enough sleep. Choices **b** and **e** are too specific to be the passage's primary purpose. Choices **a** and **c** are not supported by the passage.

172. **d.** Although he was *a man of no formal scientific education* (line 2), Leeuwenhoek demonstrated, in his own words, *a craving after knowledge, which I notice resides in me more than in most other men* (lines 24–26), who was the first to describe microorganisms. The phrase *stumbled upon* in choice **a** is too accidental to describe Leeuwenhoek's perseverance. The words *proficient* and *entertainment* in choice **c** do not accurately describe Leeuwenhoek's skill and drive depicted in the passage. Choices **b** and **e** are incorrect; Leeuwenhoek was not trained nor did he know that his discoveries would later help to cure disease.

173. **c.** *Inspired* means to exert an animating or enlivening influence on. In the context of the passage, Leeuwenhoek's creation of microscope lenses were influenced by the lenses used by drapers.

174. **a.** The quotation highlights the value Leeuwenhoek placed on sharing his discoveries with other scientists. He states that he *thought it was my duty to put down my discovery on paper, so that all ingenious people might be informed thereof* (lines 27–28).

175. **b.** The tone of the passage is positive. However, *ecstatic reverence* (choice **a**) is too positive and *tepid approval* (choice **c**) is not positive enough.

176. **c.** Nowhere in the passage does the author speculate about whether teenagers can change their exercise habits.

177. **c.** One meaning of *sedentary* is settled; another meaning is doing or requiring much sitting. *Stationary*, defined as fixed in a course or mode, is closest in meaning.

178. **e.** The last sentence illustrates factors that motivate teenagers to exercise by using the results of a national survey to provide specific examples.

179. **d.** The passage promotes change in teenagers' exercise habits by emphasizing the benefits of exercise, the moderate amount of exercise needed to achieve benefits, and some factors that may encourage teenagers to exercise.

180. **c.** The focus of the passage is Blackwell's efforts to open the profession of medicine to women. Lines 3–4 state that Blackwell *supported women's medical education and helped many other women's careers.*

181. **a.** In this context, the word *practical* refers to the solution's utility as opposed to its theoretical or ideal premise.

182. **b.** The author suggests that Samuel Blackwell's belief in slaves' rights influenced Elizabeth's struggle for greater rights for women.

183. **e.** Blackwell wrote that the study of medicne *filled me with disgust* (line 22).

184. **b.** Although Blackwell did overcome her revulsion of the body, provide healthcare to the poor, and establish a medical college for women, she did not chose medicine for these reasons according to the passage. Blackwell was told her goal was *impossible* (line 29), *too expensive* (line 29) and that medical education was *not available to women* (line 30).

185. e. The Geneva Medical College student body voted "yes" on Blackwell's admittance *as a joke* (line 38).

186. b. Initially Blackwell was interested in teaching (line 23). Subsequently, she *was attracted by the challenge* (lines 31–32) and determined to succeed in studying and practicing medicine.

187. b. The question calls for an opinion. The passage does not speculate about what degree of women's success can be attributed to Blackwell's influence.

188. b. The author uses the phrase *going gray* (line 1) as a metaphor for growing older. It describes the phenomenon of a large segment of a population growing older.

189. c. The passage makes an argument for more geriatric training based on statistical information and studies.

190. d. The passage emphasizes the need for age-specific care.

191. a. In this context, *address* most nearly means *manage*, or treat. The sentence implies that some kind of action is taken after the problem has first been identified, analyzed, and dissected.

192. c. Although choices **a** and **b** may be correct statements, they do not reflect the author's purpose in citing the example of untreated depression in the elderly. Choice **d** is incorrect and choice **e** is not supported by the passage.

193. c. According to the passage, geriatric training improves a healthcare provider's ability to *distinguish between "normal" characteristics associated with aging and illness* (lines 58–59).

194. b. The author states that healthcare providers should consider *not only the physical but mental, emotional, and social changes of the aging process* (lines 57–58).

195. d. The author's sense of urgent recommendation is expressed through the use of the helping verbs *must* (lines 55 and 56) and *should* (line 60).

196. d. Choices **a, b,** and **e** are too specific to be the primary purpose of the passage, whereas choice **c** is too general. The passage focuses on the importance of the first *official report* (line 27) to name smoking a serious health hazard.

197. c. One meaning of *preoccupied* is lost in thought; another is engaged or engrossed. In this case, *absorbed* is nearest in meaning.

198. a. The *debate over the hazards and benefits of smoking* (line 6) that continued since the *sixteenth century* (line 9) points to a long-standing controversy.

199. e. *An alliance of prominent private health organizations* (line 28) gave the push for an official report on smoking.

200. d. The quotation illustrates the response to the report, describing its effect on the country as a *bombshell* (line 48).

201. **e.** *Hedged* (line 63) can mean hindered or hemmed in, but in this instance, it most nearly means evaded. The author suggests in lines 62–67 that the report evaded a risk by calling smoking a habit rather than an addiction.

202. **b.** The author's statement implies that the report could have suggested specific actions to confront the health problem of smoking, but that it did not.

203. **b.** The author describes the influence of the report in positive terms except to mention that it did not give recommendations for remedial actions.

204. **c.** An *adage* is a word used to describe a common observation or saying, like *beauty is only skin deep* (Passage 2, line 1).

205. **c.** The author states that *the health risk for cosmetic procedures is low* (Passage 1, lines 24–25) but does not give factual information to back this claim. The statement is important to the author's argument because he or she cites it as one of the reasons his or her attitude toward plastic surgery has changed.

206. **e.** The author describes cosmetic plastic surgeons as *slick salespeople reaping large financial rewards from others' insecurity and vanity* (Passage 1, lines 17–18).

207. **d.** The author of Passage 1 directly invokes the audience he or she hopes to reach in line 31: *members of the medical community*.

208. **d.** One definition of *saturate* is to satisfy fully; another definition, which fits the context of the passage, is to fill completely with something that permeates or pervades.

209. **b.** The author of Passage 2 claims that she grew up in *the spirit of feminism* (lines 10–11), *believ[ing] that women should be valued for who they are and what they do, not for how they look* (lines 12–13). The author implies that this is a belief held by feminists of the 1970s.

210. **a.** The author of Passage 1, a physician, discusses his or her professional jealousy in lines 14–21. The author of Passage 2 does not raise this issue.

211. **d.** Passage 1 states that the demand for cosmetic surgery has increased in part *because the job market has become more competitive* (line 6). Passage 2 comments on *a competitive culture where looks count* (line 24).

212. **a.** Both passages are first-person accounts that use personal experience to build an argument.

213. **c.** Choice **d** is true, but too specific to be the author's primary purpose. Choice **e** can be eliminated because it is too negative and choices **a** and **b** are too positive.

214. **a.** The author contrasts the public's dismissal of the arcane practice of wearing garlic with its increasing acceptance of herbal remedies.

215. **b.** In this context, *conventional* refers to the established system of Western medicine or biomedicine.

216. **d.** Choice **a** is overly general and choice **b** is too negative to be inferred from the survey's findings. Choice **c** is incorrect—the author does not mention the "baby boom" age group, but that does not imply that the survey does not include it. The survey does not support the prediction in choice **e**.

217. **a.** The statistic illustrates the popularity of alternative therapies without giving any specific information as to why.

218. **e.** The author states that Americans are not replacing conventional healthcare but are adding to or supplementing it with alternative care.

219. **d.** The shortcomings of conventional healthcare mentioned in lines 30–35 are the *time constraints of managed care* (line 31), *focus on technology* (line 32), and inability to *relieve symptoms associated with chronic disease* (line 34).

220. **a.** The author states that once *scientific investigation has confirmed their safety and efficacy* (lines 37–38), alternative therapies may be accepted by the medical establishment.

221. **b.** The author gives evidence of observational studies to show that garlic may be beneficial. Choice **d** is incorrect, however, because the author emphasizes that *these findings have not been confirmed in clinical studies* (lines 51–52).

222. **d.** The passage does not offer a criticism or argument about alternative healthcare, but rather reports on the phenomenon with some playfulness.

223. **a.** The article raises the question, *Could the dietary recommendations of the last twenty years be wrong?* (lines 10–11).

224. **d.** The author expresses her objection by depicting the medical experts as extreme, *ridicul[ing]* (line 2) one diet while *extolling* (line 3) another.

225. **c.** Choices **a** and **d** are alternate definitions that do not apply to the passage. The author uses *gospel* (line 8) with its religious implications as an ironic statement, implying that scientists accepted a premise based on faith instead of on evidence.

226. **e.** The author begins with *Fact* (line 14) to introduce and highlight statistical information. She or he does not speculate about the meaning of the statistics until the next paragraph.

227. **a.** The author names a *sedentary lifestyle of TV watching and Internet surfing* (lines 24–25) as a contributing factor to the rise in obesity rates.

228. **b.** The passage suggests that the 1979 dietary guidelines responded to *a theory that dietary fat* (line 34) increases heart disease.

229. **b.** The passage describes the anti-fat message as *oversimplified* (lines 48–49) and goes on to cite the importance of certain beneficial types of fat found in olive oil and nuts (lines 38–39).

230. **c.** This example supports the claim that the body uses refined carbohydrates in *much the same way* (lines 42–43) that it does sweets.

231. **e.** Lines 42–43 support this statement.

232. **d.** The last sentence is ironic—it expresses an incongruity between conflicting dietary advice that targets different types of food as unhealthy, and the reality that humans need to eat.

Literature and Literary Criticism

Questions 233–237 are based on the following passage.

The following passage is from Frank McCourt's 1996 memoir **Angela's Ashes.** *The author describes what it was like to go to school as a young boy.*

(1)   We go to school through lanes and back streets so that we won't meet the respectable boys who go to the Christian Brothers' School or the rich ones who go to the Jesuit school, Crescent College. The Christian Brothers' boys wear tweed jackets, warm woolen sweaters, shirts,

(5)   ties, and shiny new boots. We know they're the ones who will get jobs in the civil service and help the people who run the world. The Crescent College boys wear blazers and school scarves tossed around their necks and over their shoulders to show they're cock o' the walk. They have long hair which falls across their foreheads and over their eyes so

(10)   that they can toss their quaffs like Englishmen. We know they're the ones who will go to university, take over the family business, run the government, run the world. We'll be the messenger boys on bicycles who deliver their groceries or we'll go to England to work on the building sites. Our sisters will mind their children and scrub their

(15)   floors unless they go off to England, too. We know that. We're ashamed of the way we look and if boys from the rich schools pass remarks we'll get into a fight and wind up with bloody noses or torn clothes. Our masters will have no patience with us and our fights

because their sons go to the rich schools and, Ye have no right to raise
(20)   your hands to a better class of people so ye don't.

233. The "we" the author uses throughout the passage refers to
   a. his family.
   b. the poor children in his neighborhood.
   c. the children who attend rich schools.
   d. the author and his brother.
   e. the reader and writer.

234. The passage suggests that the author goes to school
   a. in shabby clothing.
   b. in a taxi cab.
   c. in warm sweaters and shorts.
   d. on a bicycle.
   e. to become a civil servant.

235. The word *pass* as used in line 16 means to
   a. move ahead of.
   b. go by without stopping.
   c. be approved or adopted.
   d. utter.
   e. come to an end.

236. The author quotes his school masters saying *Ye have no right to raise
   your hands to a better class of people so ye don't* (lines 19–20) in order to
   a. demonstrate how strict his school masters were.
   b. contrast his school to the Christian Brothers' School and Cres-
      cent College.
   c. show how his teachers reinforced class lines.
   d. prove that the author was meant for greater things.
   e. show how people talked.

237. The passage implies that
   a. the author was determined to go to England.
   b. the author was determined to be someone who will run the
      world.
   c. the author often got into fights.
   d. the author didn't understand the idea of class and rank in
      society.
   e. one's class determined one's future.

Questions 238–242 are based on the following passage.

In this excerpt from Toni Morrison's 1970 novel The Bluest Eye, *Pauline tries to ease her loneliness by going to the movies.*

(1) One winter Pauline discovered she was pregnant. When she told Cholly, he surprised her by being pleased. [. . .] They eased back into a relationship more like the early days of their marriage, when he asked if she were tired or wanted him to bring her something from the
(5) store. In this state of ease, Pauline stopped doing day work and returned to her own housekeeping. But the loneliness in those two rooms had not gone away. When the winter sun hit the peeling green paint of the kitchen chairs, when the smoked hocks were boiling in the pot, when all she could hear was the truck delivering furniture down-
(10) stairs, she thought about back home, about how she had been all alone most of the time then too, but this lonesomeness was different. Then she stopped staring at the green chairs, at the delivery truck; she went to the movies instead. There in the dark her memory was refreshed, and she succumbed to her earlier dreams. Along with the idea of
(15) romantic love, she was introduced to another—physical beauty. Probably the most destructive ideas in the history of human thought. Both originated in envy, thrived in insecurity, and ended in disillusion.

238. Pauline and Cholly live
   a. in a two-room apartment above a store.
   b. in a delivery truck.
   c. next to a movie theater.
   d. with Pauline's family.
   e. in a housekeeper's quarters.

239. Lines 1–5 suggest that just prior to Pauline's pregnancy, Cholly had
   a. loved Pauline dearly.
   b. begun to neglect Pauline.
   c. worked every day of the week.
   d. cared about Pauline's dreams.
   e. graduated from college.

240. Pauline's loneliness is *different* from the loneliness she felt back home (lines 10–11) because
   a. she's more bored than lonely.
   b. her family has abandoned her.
   c. she wants Cholly to be more romantic.
   d. she's a mother now.
   e. she shouldn't feel lonely with Cholly.

241. Pauline's *earlier dreams* (line 14) were of
   a. romance.
   b. being beautiful.
   c. having many children.
   d. being a famous actress.
   e. owning her own store.

242. The passage suggests that going to the movies will
   a. inspire Pauline to become an actress.
   b. inspire Pauline to demand more respect from Cholly.
   c. only make Pauline more unhappy with her life.
   d. encourage Pauline to study history.
   e. create a financial strain on the family.

Questions 243–248 are based on the following passage.

In this excerpt from Sherman Alexie's novel **Reservation Blues,** *Thomas struggles with his feelings about his father, Samuel.*

(1)   Thomas, Chess, and Checkers stayed quiet for a long time. After a while, Chess and Checkers started to sing a Flathead song of mourning. *For a wake, for a wake.* Samuel was still alive, but Thomas sang along without hesitation. That mourning song was B-7 on every reser-
(5)   vation jukebox.

   After the song, Thomas stood and walked away from the table where his father lay flat as a paper plate. He walked outside and cried. Not because he needed to be alone; not because he was afraid to cry in front of women. He just wanted his tears to be individual, not tribal.
(10)   Those tribal tears collected and fermented in huge BIA [Bureau of Indian Affairs] barrels. Then the BIA poured those tears into beer and Pepsi cans and distributed them back onto the reservation. Thomas wanted his tears to be selfish and fresh.

   "Hello," he said to the night sky. He wanted to say the first word of
(15)   a prayer or a joke. A prayer or a joke often sound alike on the reservation.

"Help," he said to the ground. He knew the words to a million songs: Indian, European, African, Mexican, Asian. He sang "Stairway to Heaven" in four different languages but never knew where that

(20) staircase stood. He sang the same Indian songs continually but never sang them correctly. He wanted to make his guitar sound like a waterfall, like a spear striking salmon, but his guitar only sounded like a guitar. He wanted the songs, the stories, to save everybody.

243. Thomas, Chess, and Checkers are
  a. Mexican.
  b. European.
  c. Asian.
  d. African.
  e. Native American.

244. In line 3, *a wake* means
  a. the turbulence left behind by something moving through water.
  b. no longer asleep.
  c. a viewing of a dead person before burial.
  d. aftermath.
  e. celebration.

245. The fact that Thomas, Chess, and Checkers sing a song of mourning while Samuel is still alive suggests that
  a. Samuel is afraid to die.
  b. Samuel doesn't belong on the reservation.
  c. Samuel's life is tragic.
  d. they believe the song has healing powers.
  e. Samuel is a ghost.

246. Thomas wants his tears to be "selfish and fresh" (line 13) because
  a. it is difficult for him to share his feelings with others.
  b. he wants to mourn his father as an individual, not just as another dying Indian.
  c. he feels guilty mourning his father before his father has died.
  d. he doesn't think the tribe will mourn his father's passing.
  e. tribal tears were meaningless.

247. The sentence *Then the BIA poured those tears into beer and Pepsi cans and distributed them back onto the reservation* (lines 11–12) is an example of
a. a paradox.
b. dramatic irony.
c. onomatopoeia.
d. flashback.
e. figurative language.

248. In line 17, Thomas asks for help because
a. he can't stop crying.
b. he wants to be a better guitar player.
c. he wants to be able to rescue people with his music.
d. he can't remember the words to the song.
e. no one wants to listen to him play.

Questions 249–256 are based on the following passage.

In this excerpt from John Steinbeck's 1936 novel In Dubious Battle, *Mac and Doc Burton discuss "the cause" that leads hundreds of migratory farm workers to unite and strike against landowners.*

(1)   Mac spoke softly, for the night seemed to be listening. "You're a mystery to me, too, Doc."

"Me? A mystery?"

"Yes, you. You're not a Party man, but you work with us all the time;
(5)   you never get anything for it. I don't know whether you believe in what we're doing or not, you never say, you just work. I've been out with you before, and I'm not sure you believe in the cause at all."

Dr. Burton laughed softly. "It would be hard to say. I could tell you some of the things I think; you might not like them. I'm pretty sure you
(10)   won't like them."

"Well, let's hear them anyway."

"Well, you say I don't believe in the cause. That's not like not believing in the moon. There've been communes before, and there will be again. But you people have an idea that if you can *establish* the thing, the
(15)   job'll be done. Nothing stops, Mac. If you were able to put an idea into effect tomorrow, it would start changing right away. Establish a commune, and the same gradual flux will continue."

"Then you don't think the cause is good?"

Burton sighed. "You see? We're going to pile up on that old rock
(20)   again. That's why I don't like to talk very often. Listen to me, Mac. My senses aren't above reproach, but they're all I have. I want to see the

whole picture—as nearly as I can. I don't want to put on the blinders of 'good' and 'bad,' and limit my vision. If I used the term 'good' on a thing I'd lose my license to inspect it, because there might be bad in it. Don't
(25) you see? I want to be able to look at the whole thing."

Mac broke in heatedly, "How about social injustice? The profit system? You have to say they're bad."

Dr. Burton threw back his head and looked at the sky. "Mac," he said. "Look at the physiological injustice, the injustice of tetanus [. . .], the
(30) gangster methods of amoebic dysentery—that's my field."

"Revolution and communism will cure social injustice."

"Yes, and disinfection and prophylaxis will prevent others."

"It's different, though; men are doing one, and germs are doing the other."
(35) "I can't see much difference, Mac."

[. . .] "Why do you hang around with us if you aren't for us?"

"I want to *see*," Burton said. "When you cut your finger, and streptococci get in the wound, there's a swelling and a soreness. That swelling is the fight your body puts up, the pain is the battle. You can't tell which
(40) one is going to win, but the wound is the first battleground. If the cells lose the first fight the streptococci invade, and the fight goes on up the arm. Mac, these little strikes are like the infection. Something has got into the men; a little fever has started and the lymphatic glands are shooting in the reinforcements. I want to see, so I go to the seat of the wound."
(45) "You figure the strike is a wound?"

"Yes. Group-men are always getting some kind of infection. This seems to be a bad one. I want to *see*, Mac. I want to watch these group-men, for they seem to me to be a new individual, not at all like single men. A man in a group isn't himself at all, he's a cell in
(50) an organism that isn't like him any more than the cells in your body are like you. I want to watch the group, and see what it's like. People have said, 'mobs are crazy, you can't tell what they'll do.' Why don't people look at mobs not as men, but as mobs? A mob nearly always seems to act reasonably, for a mob."
(55) "Well, what's this got to do with the cause?"

"It might be like this, Mac: When group-man wants to move, he makes a standard. 'God wills that we recapture the Holy Land'; or he says, 'We fight to make the world safe for democracy'; or he says, 'We will wipe out social injustice with communism.' But the group doesn't care about the
(60) Holy Land, or Democracy, or Communism. Maybe the group simply wants to move, to fight, and uses these words simply to reassure the brains of individual men. I say it *might* be like that, Mac."

"Not with the cause, it isn't," Mac cried.

249. In lines 15–17, Doc Burton argues that
  a. even if the cause succeeds, it won't change anything.
  b. the cause is unstoppable.
  c. the supporters of the cause should establish a commune.
  d. the cause itself is always changing.
  e. change can only come about gradually.

250. The *cause* the men refer to throughout the passage is
  a. democracy.
  b. communism.
  c. capitalism.
  d. insurgency.
  e. freedom.

251. Doc Burton is best described as
  a. an objective observer.
  b. a representative of the government.
  c. a staunch supporter of the cause.
  d. a visionary leader.
  e. a reluctant participant.

252. According to Doc Burton, the *strikes are like the infection* (line 42) because
  a. the strikes are life-threatening.
  b. many of the strikers are ill.
  c. the size of the group has swollen.
  d. the strikes are a reaction to an injury.
  e. the strikes are taking place on a battleground.

253. By comparing *group-men* to a living organism (lines 48–50), Doc Burton
  a. reinforces his idea that individuals are lost in the larger whole.
  b. shows that group-men is constantly changing and growing.
  c. supports his assertion that the strikers are like an infection.
  d. explains why he is with the strikers.
  e. reflects his opinion that the strikes' success depends upon unity within the group.

254. According to Doc Burton, the main difference between *group-men* and the individual is that
   a. individuals can be controlled but groups cannot.
   b. individuals do not want to fight but groups do.
   c. individuals may believe in a cause but groups do not.
   d. groups are often crazy but individuals are not.
   e. people in groups can reassure one another.

255. It can be inferred from this passage that Doc Burton believes the cause
   a. is just an excuse for fighting.
   b. is reasonable.
   c. will fail.
   d. will correct social injustice.
   e. will make America a more democratic place.

256. Doc Burton repeats the word *might* in lines 56 and 62 because
   a. he doesn't believe Mac is sincere about the cause.
   b. he really wants Mac to consider the possibility that the group is blind to the cause.
   c. he is asking a rhetorical question.
   d. he doesn't want Mac to know the truth about the cause.
   e. he wants Mac to see that he isn't really serious in his criticism of the cause.

Questions 257–265 are based on the following passage.

In this passage, written in 1925, writer Edith Wharton distinguishes between subjects suitable for short stories and those suitable for novels.

(1)   It is sometimes said that a "good subject" for a short story should always be capable of being expanded into a novel.

The principle may be defendable in special cases; but it is certainly a misleading one on which to build any general theory. Every "subject"
(5)   (in the novelist's sense of the term) must necessarily contain within itself its own dimensions; and one of the fiction-writer's essential gifts is that of discerning whether the subject which presents itself to him, asking for incarnation, is suited to the proportions of a short story or of a novel. If it appears to be adapted to both the chances are that it is
(10)   inadequate to either.

It would be a great mistake, however, to try to base a hard-and-fast theory on the denial of the rule as on its assertion. Instances of short stories made out of subjects that could have been expanded into a

(15) novel, and that are yet typical short stories and not mere stunted novels, will occur to everyone. General rules in art are useful chiefly as a lamp in a mine, or a handrail down a black stairway; they are necessary for the sake of the guidance they give, but it is a mistake, once they are formulated, to be too much in awe of them.

(20) There are at least two reasons why a subject should find expression in novel-form rather than as a tale; but neither is based on the number of what may be conveniently called incidents, or external happenings, which the narrative contains. There are novels of action which might be condensed into short stories without the loss of their distinguishing qualities. The marks of the subject requiring a longer

(25) development are, first, the gradual unfolding of the inner life of its characters, and secondly the need of producing in the reader's mind the sense of the lapse of time. Outward events of the most varied and exciting nature may without loss of probability be crowded into a few hours, but moral dramas usually have their roots deep in the soul, their rise far

(30) back in time; and the suddenest-seeming clash in which they culminate should be led up to step by step if it is to explain and justify itself.

There are cases, indeed, when the short story may make use of the moral drama at its culmination. If the incident dealt with be one which a single retrospective flash sufficiently lights up, it is qualified for use

(35) as a short story; but if the subject be so complex, and its successive phases so interesting, as to justify elaboration, the lapse of time must necessarily be suggested, and the novel-form becomes appropriate.

The effect of compactness and instantaneity sought in the short story is attained mainly by the observance of two "unities"—the old

(40) traditional one of time, and that other, more modern and complex, which requires that any rapidly enacted episode shall be seen through only one pair of eyes . . . .

One thing more is needful for the ultimate effect of probability; and that is, never let the character who serves as reflector record anything

(45) not naturally within his register. It should be the storyteller's first care to choose this reflecting mind deliberately, as one would choose a building-site, or decide upon the orientation of one's house, and when this is done, to live inside the mind chosen, trying to feel, see and react exactly as the latter would, no more, no less, and, above all, no other-

(50) wise. Only thus can the writer avoid attributing incongruities of thought and metaphor to his chosen interpreter.

257. In the opening sentence (lines 1–2), the author
   a. states her main idea.
   b. states the idea she will disprove.
   c. presents an example of the point she will prove.
   d. presents an anecdote to capture the reader's attention.
   e. presents evidence for her thesis.

258. The author's main purpose in this passage is to
   a. provide guidelines for choosing the narrator in a novel.
   b. provide tips for making short stories and novels more realistic.
   c. debunk several myths about writing novels.
   d. explain why some tales are better for novels than short stories.
   e. provide strategies for writers to develop ideas for short stories
     and novels.

259. The author believes that rules for writing
   a. should always be strictly adhered to.
   b. should only be general guidelines.
   c. should be revised regularly.
   d. are just good common sense.
   e. are too theoretical.

260. In lines 15–18 the author uses
   a. analogy.
   b. personification.
   c. hyperbole.
   d. foreshadowing.
   e. innuendo.

261. According to the author, which factor(s) determine whether a
   subject is suitable for a novel instead of a short story?
   I. the number of incidents in the story
   II. the need to show the development of the character(s)
   III. the need to reflect the passage of time
   a. I only
   b. I and II only
   c. II and III only
   d. I and III only
   e. all of the above

262. In lines 32–37, the author
    a. contradicts the rule established in the previous paragraph.
    b. clarifies the rule established in the previous paragraph.
    c. shows an example of the rule established in the previous paragraph.
    d. justifies the rule established in the previous paragraph.
    e. provides a new rule.

263. According to the author, two defining characteristics of a short story are
    a. complexity and probability.
    b. moral dilemmas and sudden clashes.
    c. retrospection and justification.
    d. metaphor and congruity.
    e. limited time and point of view.

264. In line 46, *this reflecting mind* refers to
    a. the author.
    b. the narrator.
    c. the reader.
    d. a story's translator.
    e. a story's editor.

Questions 265–273 are based on the following passage.

This excerpt is from the final scene of the play George Bernard Shaw's 1916 play Pygmalion, *when Professor Higgins learns just how well he taught Liza.*

(1)   HIGGINS: If you're going to be a lady, you'll have to give up feeling neglected if the men you know don't spend half their time sniveling over you and the other half giving you black eyes. If you can't stand the coldness of my sort of life, and the strain of it, go back to
(5)   the gutter. Work 'til you are more a brute than a human being; and then cuddle and squabble and drink 'til you fall asleep. Oh, it's a fine life, the life of the gutter. It's real: it's warm: it's violent: you can feel it through the thickest skin: you can taste it and smell it without any training or any work. Not like Science and Literature and Classi-
(10)   cal Music and Philosophy and Art. You find me cold, unfeeling, self-ish, don't you? Very well: be off with you to the sort of people you like. Marry some sentimental hog or other with lots of money, and a thick pair of lips to kiss you with and a thick pair of boots to kick you with. If you can't appreciate what you've got, you'd better get
(15)   what you can appreciate.

LIZA (*desperate*): Oh, you are a cruel tyrant. I can't talk to you: you
turn everything against me: I'm always in the wrong. But you know
very well all the time that you're nothing but a bully. You know I
can't go back to the gutter, as you call it, and that I have no real
(20) friends in the world but you and the Colonel. You know well I
couldn't bear to live with a low common man after you two; and it's
wicked and cruel of you to insult me by pretending I could. You
think I must go back to Wimpole Street because I have nowhere
else to go but father's. But don't you be too sure that you have me
(25) under your feet to be trampled on and talked down. I'll marry
Freddy, I will, as soon as he's able to support me.
HIGGINS (*sitting down beside her*): Rubbish! You shall marry an
ambassador. You shall marry the Governor-General of India or the
Lord-Lieutenant of Ireland, or somebody who wants a deputy-
(30) queen. I'm not going to have my masterpiece thrown away on
Freddy.
LIZA: You think I like you to say that. But I haven't forgot what you
said a minute ago; and I won't be coaxed round as if I was a baby or
a puppy. If I can't have kindness, I'll have independence.
(35) HIGGINS: Independence? That's middle class blasphemy. We are all
dependent on one another, every soul of us on earth.
LIZA (*rising determinedly*): I'll let you see whether I'm dependent on
you. If you can preach, I can teach. I'll go and be a teacher.
HIGGINS: What'll you teach, in heaven's name?
(40) LIZA: What you taught me. I'll teach phonetics.
HIGGINS: Ha! ha! ha!
LIZA: I'll offer myself as an assistant to Professor Nepean.
HIGGINS (*rising in a fury*): What! That impostor! that humbug! that
toadying ignoramus! Teach him my methods! my discoveries! You
(45) take one step in his direction and I'll wring your neck. (*He lays hands
on her.*) Do you hear?
LIZA (*defiantly resistant*): Wring away. What do I care? I knew you'd
strike me some day. (*He lets her go, stamping with rage at having for-
gotten himself, and recoils so hastily that he stumbles back into his seat on
(50) the ottoman.*) Aha! Now I know how to deal with you. What a fool
I was not to think of it before! You can't take away the knowledge
you gave me. You said I had a finer ear than you. And I can be civil
and kind to people, which is more than you can. Aha! That's done
you, Henry Higgins, it has. Now I don't care that (*snapping her fin-
(55) gers*) for your bullying and your big talk. I'll advertise it in the
papers that your duchess is only a flower girl that you taught, and
that she'll teach anybody to be a duchess just the same in six months

for a thousand guineas. Oh, when I think of myself crawling under your feet and being trampled on and called names, when all the time I had only to lift up my finger to be as good as you, I could just kick myself.

(60)

265. In lines 1–15, Higgins contrasts the *life of the gutter* with his *sort of life*, which is best described as
   a. the life of an ambassador.
   b. the life of the rich and famous.
   c. the life of a tyrant.
   d. the life of a scholar.
   e. the life of the working class.

266. Wimpole Street (line 23) is most likely
   a. a fashionable area.
   b. where Professor Nepean resides.
   c. where Higgins teaches.
   d. where Freddy lives.
   e. where Liza grew up.

267. Liza wants Higgins to
   a. appreciate her work.
   b. help her find a suitable husband.
   c. marry her.
   d. teach her everything he knows.
   e. treat her with more respect.

268. The word *common* in line 21 means
   a. usual.
   b. unrefined.
   c. popular.
   d. average.
   e. shared by two or more.

269. In lines 43–46, Higgins proves that
   a. he is a bully.
   b. Liza can't teach with Professor Nepean.
   c. Professor Nepean is a fake.
   d. he and Liza depend upon each other.
   e. he knows better than Liza.

270. Higgins' use of the word *masterpiece* in line 30 implies that
   a. he is an artist.
   b. he thinks Liza is very beautiful.
   c. he thinks of Liza as his creation.
   d. he is in love with Liza.
   e. Liza is his servant.

271. Which of the following best describes what Higgins has taught Liza?
   a. the history of the English language.
   b. how to speak and act like someone from the upper class.
   c. how to be independent of others.
   d. how to understand literature and philosophy.
   e. how to appreciate scholarly work.

272. In lines 37–61, the main reason Higgins is so upset is because
   a. Liza threatens to teach his methods to others.
   b. he realizes he has been a bad teacher.
   c. he realizes he is as abusive as someone from *the gutter.*
   d. he realizes he cannot control Liza.
   e. he realizes Liza does not love him anymore.

273. The passage implies that Liza's most significant transformation in the play is from
   a. lower class to upper class.
   b. ignorant to educated.
   c. oppressed to empowered.
   d. single to married.
   e. cold to compassionate.

Questions 274–281 are based on the following passage.

In this excerpt from Charlotte Bronte's novel Jane Eyre, *the narrator decides to leave Lowood, the boarding school where she has lived for eight years.*

(1)   Miss Temple, through all changes, had thus far continued superintendent of the seminary; to her instruction I owed the best part of my acquirements; her friendship and society had been my continual solace: she had stood me in the stead of mother, governess, and, latterly,
(5)   companion. At this period she married, removed with her husband (a clergyman, an excellent man, almost worthy of such a wife) to a distant county, and consequently was lost to me.

From the day she left I was no longer the same: with her was gone every settled feeling, every association that had made Lowood in some

(10) degree a home to me. I had imbibed from her something of her nature and much of her habits: more harmonious thoughts: what seemed better-regulated feelings had become inmates of my mind. I had given in allegiance to duty and order; I was quiet; I believed I was content: to the eyes of others, usually even to my own, I appeared a disciplined and sub-

(15) dued character.

But destiny, in the shape of the Rev. Mr. Nasmyth, came between me and Miss Temple: I saw her in her traveling dress step into a post-chaise, shortly after the marriage ceremony; I watched the chaise mount the hill and disappear beyond its brow; and then retired to my own room, and

(20) there spent in solitude the greatest part of the half-holiday granted in honor of the occasion.

I walked about the chamber most of the time. I imagined myself only to be regretting my loss, and thinking how to repair it; but when my reflections concluded, and I looked up and found that the afternoon was

(25) gone, and evening far advanced, another discovery dawned on me, namely, that in the interval I had undergone a transforming process; that my mind had put off all it had borrowed of Miss Temple—or rather that she had taken with her the serene atmosphere I had been breathing in her vicinity—and that now I was left in my natural element, and begin-

(30) ning to feel the stirring of old emotions. It did not seem as if a prop were withdrawn, but rather as if a motive were gone; it was not the power to be tranquil which had failed me, but the reason for tranquility was no more. My world had for some years been in Lowood: my experience had been of its rules and systems; now I remembered that the real world

(35) was wide, and that a varied field of hopes and fears, of sensations and excitements, awaited those who had courage to go forth into its expanse, to seek real knowledge of life amidst its perils.

I went to my window, opened it, and looked out. There were the two wings of the building; there was the garden; there were the skirts of

(40) Lowood; there was the hilly horizon. My eye passed all other objects to rest on those most remote, the blue peaks: it was those I longed to surmount; all within their boundary of rock and heath seemed prison-ground, exile limits. I traced the white road winding round the base of one mountain, and vanishing in a gorge between two: how I longed to

(45) follow it further! I recalled the time when I had traveled that very road in a coach; I remembered descending that hill at twilight: an age seemed to have elapsed since the day which brought me first to Lowood, and I had never quitted it since. My vacations had all been spent at school: Mrs. Reed had never sent for me to Gateshead; neither she nor any of

(50) her family had ever been to visit me. I had had no communication by
letter or message with the outer world: school-rules, school-duties,
school-habits and notions, and voices, and faces, and phrases, and cos-
tumes, and preferences, and antipathies: such was what I knew of existence. And now I felt that it was not enough: I tired of the routine of
(55) eight years in one afternoon. I desired liberty; for liberty I gasped; for
liberty I uttered a prayer; it seemed scattered on the wind then faintly
blowing. I abandoned it and framed a humbler supplication; for change,
stimulus: that petition, too, seemed swept off into vague space: "Then,"
I cried, half desperate, "grant me at least a new servitude!"

274. Miss Temple was the narrator's
 I. teacher.
 II. friend.
 III. mother.
 a. I only
 b. II only
 c. III only
 d. I and II
 e. all of the above

275. While Miss Temple was at Lowood, the narrator
 a. was calm and content.
 b. was often alone.
 c. had frequent disciplinary problems.
 d. longed to leave Lowood.
 e. felt as if she were in a prison.

276. The word *inmates* in line 12 means
 a. captives.
 b. patients.
 c. prisoners.
 d. residents.
 e. convalescents.

277. Mrs. Reed (line 49) is most likely
 a. the narrator's mother.
 b. the head mistress of Lowood.
 c. the narrator's former guardian.
 d. the narrator's friend.
 e. a fellow student at Lowood.

278. It can be inferred from the passage that life at Lowood was
   a. very unconventional and modern.
   b. very structured and isolated.
   c. harsh and demeaning.
   d. liberal and carefree.
   e. urban and sophisticated.

279. After Miss Temple's wedding, the narrator
   a. realizes she wants to experience the world.
   b. decides that she must get married.
   c. realizes she can never leave Lowood.
   d. decides to return to her family at Gateshead.
   e. determines to follow Miss Temple.

280. The passage suggests that the narrator
   a. will soon return to Lowood.
   b. was sent to Lowood by mistake.
   c. is entirely dependent upon Miss Temple.
   d. has run away from Lowood before.
   e. is naturally curious and rebellious.

281. In lines 60–66, the narrator reduces her petition to simply a *new servitude* because she
   a. doesn't believe in prayer.
   b. is not in a free country.
   c. has been offered a position as a servant.
   d. knows so little of the real world.
   e. has been treated like a slave at Lowood.

Questions 282–289 are based on the following passage.

In this excerpt from Susan Glaspell's one-act play **Trifles,** *Mrs. and Mrs. Peters make an important discovery in Mrs. Wright's home as their husbands try to determine who strangled Mr. Wright.*

(1)   MRS. PETERS: Well, I must get these things wrapped up. They may be through sooner than we think. [*Putting apron and other things together.*] I wonder where I can find a piece of paper, and string.
   MRS. HALE: In that cupboard, maybe.
(5)   MRS. PETERS [*looking in cupboard*]: Why, here's a birdcage. [*Holds it up.*] Did she have a bird, Mrs. Hale?
   MRS. HALE: Why, I don't know whether she did or not—I've not been here for so long. There was a man around last year selling

canaries cheap, but I don't know as she took one; maybe she did.
(10)   She used to sing real pretty herself.

MRS. PETERS [*glancing around*]: Seems funny to think of a bird here. But she must have had one, or why would she have a cage? I wonder what happened to it.

MRS. HALE: I s'pose maybe the cat got it.

(15)   MRS. PETERS: No, she didn't have a cat. She's got that feeling some people have about cats—being afraid of them. My cat got in her room and she was real upset and asked me to take it out.

MRS. HALE: My sister Bessie was like that. Queer, ain't it?

MRS. PETERS [*examining the cage*]: Why, look at this door. It's broke.
(20)   One hinge is pulled apart.

MRS. HALE [*looking too*]: Looks as if someone must have been rough with it.

MRS. PETERS: Why, yes.

[*She brings the cage forward and puts it on the table.*]

(25)   MRS. HALE: I wish if they're going to find any evidence they'd be about it. I don't like this place.

MRS. PETERS: But I'm awful glad you came with me, Mrs. Hale. It would be lonesome for me sitting here alone.

MRS. HALE: It would, wouldn't it? [*Dropping her sewing.*] But I tell
(30)   you what I do wish, Mrs. Peters. I wish I had come over sometimes when *she* was here. I—[*looking around the room*]—wish I had.

MRS. PETERS: But of course you were awful busy, Mrs. Hale—your house and your children.

MRS. HALE: I could've come. I stayed away because it weren't cheer-
(35)   ful—and that's why I ought to have come. I—I've never liked this place. Maybe because it's down in a hollow and you don't see the road. I dunno what it is but it's a lonesome place and always was. I wish I had come over to see Minnie Foster sometimes. I can see now—

(40)   [*Shakes her head.*]

MRS. PETERS: Well, you mustn't reproach yourself, Mrs. Hale. Somehow we just don't see how it is with other folks until—something comes up.

MRS. HALE: Not having children makes less work—but it makes a
(45)   quiet house, and Wright out to work all day, and no company when he did come in. Did you know John Wright, Mrs. Peters?

MRS. PETERS: Not to know him; I've seen him in town. They say he was a good man.

MRS. HALE: Yes—good; he didn't drink, and kept his word as well
(50)   as most, I guess, and paid his debts. But he was a hard man, Mrs.

Peters. Just to pass the time of day with him—[*shivers*]. Like a raw wind that gets to the bone. [*Pauses, her eye falling on the cage.*] I should think she would'a wanted a bird. But what do you suppose went with it?

MRS. PETERS: I don't know, unless it got sick and died.

(55)   [*She reaches over and swings the broken door, swings it again. Both women watch it.*]

MRS. HALE: You weren't raised round here, were you? [MRS. PETERS *shakes her head.*] You didn't know—her?

MRS. PETERS: Not till they brought her yesterday.

(60)   MRS. HALE: She—come to think of it, she was kind of like a bird herself—real sweet and pretty, but kind of timid and—fluttery. How—she—did—change. [*Silence; then as if struck by a happy thought and relieved to get back to every day things.*] Tell you what, Mrs. Peters, why don't you take the quilt in with you? It might take up her mind.

(65)   MRS. PETERS: Why, I think that's a real nice idea, Mrs. Hale. There couldn't possibly be any objection to it, could there? Now, just what would I take? I wonder if her patches are in here—and her things.

[*They look in the sewing basket.*]

MRS. HALE: Here's some red. I expect this has got sewing things in it.

(70)   [*Brings out a fancy box.*] What a pretty box. Looks like something somebody would give you. Maybe her scissors are in here. [*Opens box. Suddenly puts her hand to her nose.*] Why—[MRS. PETERS *bends nearer, then turns her face away.*] There's something wrapped in this piece of silk.

(75)   MRS. PETERS [*lifting the silk*]: Why this isn't her scissors.

MRS. HALE [*lifting the silk*]: Oh, Mrs. Peters—it's—

[MRS. PETERS *bends closer.*]

MRS. PETERS: It's the bird.

MRS. HALE [*jumping up*]: But, Mrs. Peters—look at it! Its neck! Look

(80)   at its neck! It's all—to the other side.

MRS. PETERS: Somebody—wrung—its—neck.

[*Their eyes meet. A look of growing comprehension, of horror. Steps are heard outside. MRS. HALE slips box under quilt pieces, and sinks into her chair. Enter SHERIFF and COUNTY ATTORNEY HALE. MRS.*

(85)   PETERS *rises.*]

282. Based on the passage, the reader can conclude that

a. Mrs. Peters and Mrs. Hale are old friends.

b. Mrs. Peters and Mrs. Hale both know Mrs. Wright very well.

c. Mrs. Peters and Mrs. Hale don't know each other very well.

d. Neither Mrs. Peters nor Mrs. Hale like Mrs. Wright.

e. Neither Mrs. Peters nor Mrs. Hale have children.

283. Mrs. Hale says she wishes she had come to Mrs. Wright's house (lines 29–31 and 37–39) because
   a. she realizes that Mrs. Wright must have been lonely.
   b. she enjoyed Mr. Wright's company.
   c. she always felt at home in the Wright's house.
   d. she realizes how important it is to keep good relationships with one's neighbors.
   e. she had a lot in common with Mrs. Wright.

284. According to Mrs. Hale, what sort of man was Mr. Wright?
   a. gentle and loving
   b. violent and abusive
   c. honest and dependable
   d. quiet and cold
   e. a strict disciplinarian

285. In lines 60–62, Mrs. Hale suggests that Mrs. Wright
   a. had become even more like a bird than before.
   b. had grown bitter and unhappy over the years.
   c. was too shy to maintain an intimate friendship.
   d. must have taken excellent care of her bird.
   e. was always singing and flitting about.

286. The phrase *take up her mind* in line 64 means
   a. worry her.
   b. make her angry.
   c. refresh her memory.
   d. keep her busy.
   e. make her think.

287. It can be inferred that Mrs. Wright
   a. got the bird as a present for her husband.
   b. was forced into marrying Mr. Wright.
   c. loved the bird because it reminded her of how she used to be.
   d. had a pet bird as a little girl.
   e. fought often with Mr. Wright.

288. When the women share a *look of growing comprehension, of horror* (line 83), they realize that
   a. Mrs. Wright killed the bird.
   b. Mr. Wright killed the bird, and Mrs. Wright killed him.
   c. they would get in trouble if the sheriff found out they were looking around in the kitchen.
   d. there's a secret message hidden in the quilt.
   e. they might be Mrs. Wright's next victims.

289. The stage directions in lines 83–86 suggest that
   a. the women are mistaken in their conclusion.
   b. the women will tell the men what they found.
   c. the women will confront Mrs. Wright.
   d. the women will keep their discovery a secret.
   e. the men had been eavesdropping on the women.

Questions 290–298 are based on the following passages.

In Passage 1, an excerpt from Mary Shelley's Frankenstein, *Victor Frankenstein explains his motive for creating his creature. In Passage 2, an excerpt from H.G. Wells' 1896 novel* The Island of Dr. Moreau, *Dr. Moreau explains to the narrator why he has been performing experiments on animals to transform them into humans.*

PASSAGE 1

(1)   I see by your eagerness, and the wonder and hope which your eyes express, my friend, that you expect to be informed of the secret with which I am acquainted; that cannot be: listen patiently until the end of my story, and you will easily perceive why I am reserved upon that
(5)   subject. I will not lead you on, unguarded and ardent as I then was, to your destruction and infallible misery. Learn from me, if not by my precepts, at least by my example, how dangerous is the acquirement of knowledge, and how much happier that man is who believes his native town to be the world, than he who aspires to become greater than his
(10) nature will allow.
       When I found so astonishing a power placed within my hands, I hesitated a long time concerning the manner in which I should employ it. Although I possessed the capacity of bestowing animation, yet to prepare a frame for the reception of it, with all its intricacies of fibers,
(15) muscles, and veins, still remained a work of inconceivable difficulty and labour. I doubted at first whether I should attempt the creation of

(20) a being like myself, or one of simpler organization; but my imagination was too much exalted by my first success to permit me to doubt of my ability to give life to an animal as complex and wonderful as man. The materials at present within my command hardly appeared adequate to so arduous an undertaking; but I doubted not that I should ultimately succeed. I prepared myself for a multitude of reverses; my operations might be incessantly baffled, and at last my work be imperfect: yet, when I considered the improvement which every day takes

(25) place in science and mechanics, I was encouraged to hope my present attempts would at least lay the foundations of future success. Nor could I consider the magnitude and complexity of my plan as any argument of its impracticability. It was with these feelings that I began the creation of my human being. As the minuteness of the parts

(30) formed a great hindrance to my speed, I resolved, contrary to my first intention, to make the being of a gigantic stature; that is to say, about eight feet in height, and proportionably large. After having formed this determination, and having spent some months in successfully collecting and arranging my materials, I began.

(35) No one can conceive the variety of feelings which bore me onwards, like a hurricane, in the first enthusiasm of success. Life and death appeared to me ideal bounds, which I should first break through, and pour a torrent of light into our dark world. A new species would bless me as its creator and source; many happy and excellent natures would

(40) owe their being to me. No father could claim the gratitude of his child so completely as I should deserve theirs. Pursuing these reflections, I thought, that if I could bestow animation upon lifeless matter, I might in process of time (although I now found it impossible) renew life where death had apparently devoted the body to corruption.

(45) These thoughts supported my spirits, while I pursued my undertaking with unremitting ardour. My cheek had grown pale with study, and my person had become emaciated with confinement. Sometimes, on the very brink of certainty, I failed; yet still I clung to the hope which the next day or the next hour might realize. One secret which

(50) I alone possessed was the hope to which I had dedicated myself; and the moon gazed on my midnight labors, while, with unrelaxed and breathless eagerness, I pursued nature to her hiding-places. Who shall conceive the horrors of my secret toil, as I dabbled among the unhallowed damps of the grave, or tortured the living animal to animate the

(55) lifeless clay? My limbs now tremble, and my eyes swim with the remembrance; but then a resistless, and almost frantic, impulse urged me forward; I seemed to have lost all soul or sensation but for this one pursuit.

PASSAGE 2

(1)   "Yes. These creatures you have seen are animals carven and wrought into new shapes. To that—to the study of the plasticity of living forms—my life has been devoted. I have studied for years, gaining in knowledge as I go. I see you look horrified, and yet I am telling you

(5)   nothing new. It all lay in the surface of practical anatomy years ago, but no one had the temerity to touch it. It's not simply the outward form of an animal I can change. The physiology, the chemical rhythm of the creature, may also be made to undergo an enduring modification, of which vaccination and other methods of inoculation with liv-

(10) ing or dead matter are examples that will, no doubt, be familiar to you.

"A similar operation is the transfusion of blood, with which subject indeed I began. These are all familiar cases. Less so, and probably far more extensive, were the operations of those medieval practitioners who made dwarfs and beggar cripples and show-monsters; some vestiges of

(15) whose art still remain in the preliminary manipulation of the young mountebank or contortionist. Victor Hugo gives an account of them in *L'Homme qui Rit.* . . . But perhaps my meaning grows plain now. You begin to see that it is a possible thing to transplant tissue from one part of an animal to another, or from one animal to another, to alter its

(20) chemical reactions and methods of growth, to modify the articulations of its limbs, and indeed to change it in its most intimate structure?

"And yet this extraordinary branch of knowledge has never been sought as an end, and systematically, by modern investigators, until I took it up! Some such things have been hit upon in the last resort of

(25) surgery; most of the kindred evidence that will recur to your mind has been demonstrated, as it were, by accident—by tyrants, by criminals, by the breeders of horses and dogs, by all kinds of untrained clumsy-handed men working for their own immediate ends. I was the first man to take up this question armed with antiseptic surgery, and with

(30) a really scientific knowledge of the laws of growth.

"Yet one would imagine it must have been practiced in secret before. Such creatures as Siamese Twins . . . . And in the vaults of the Inquisi-tion. No doubt their chief aim was artistic torture, but some, at least, of the inquisitors must have had a touch of scientific curiosity . . . ."

(35) "But," said I. "These things—these animals *talk!*"

He said that was so, and proceeded to point out that the possibili-ties of vivisection do not stop at a mere physical metamorphosis. A pig may be educated. The mental structure is even less determinate than the bodily. In our growing science of hypnotism we find the promise

(40) of a possibility of replacing old inherent instincts by new suggestions, grafting upon or replacing the inherited fixed ideas. [. . .]

But I asked him why he had taken the human form as a model. There seemed to me then, and there still seems to me now, a strange wickedness in that choice.

(45) He confessed that he had chosen that form by chance.

"I might just as well have worked to form sheep into llamas, and llamas into sheep. I suppose there is something in the human form that appeals to the artistic turn of mind more powerfully than any animal shape can. But I've not confined myself to man-making. Once or

(50) twice . . . ." He was silent, for a minute perhaps. "These years! How they have slipped by! And here I have wasted a day saving your life, and am now wasting an hour explaining myself!"

"But," said I, "I still do not understand. Where is your justification for inflicting all this pain? The only thing that could excuse vivisection

(55) to me would be some application—"

"Precisely," said he. "But you see I am differently constituted. We are on different platforms. You are a materialist."

"I am *not* a materialist," I began hotly.

"In my view—in my view. For it is just this question of pain that

(60) parts us. So long as visible or audible pain turns you sick, so long as your own pain drives you, so long as pain underlies your propositions about sin, so long, I tell you, you are an animal, thinking a little less obscurely what an animal feels. This pain—"

I gave an impatient shrug at such sophistry.

(65) "Oh! But it is such a little thing. A mind truly open to what science has to teach must see that it is a little thing."

290. In the first paragraph of Passage 1 (lines 1–10), Frankenstein reveals that the purpose of his tale is to
   a. entertain the reader.
   b. explain a scientific principle.
   c. teach a moral lesson.
   d. share the secret of his research.
   c. reveal his true nature.

291. The word *baffled* in line 23 means
   a. hindered.
   b. confused.
   c. puzzled.
   d. eluded.
   e. regulated.

292. During the creation process, Frankenstein could best be
described as
a. calm.
b. horrified.
c. evil.
d. indifferent.
e. obsessed.

293. From Passage 2, it can be inferred that Dr. Moreau is what sort of
scientist?
a. artistic
b. calculating and systematic
c. careless, haphazard
d. famous, renowned
e. materialist

294. *These things* that the narrator refers to in Passage 2, line 35 are
a. Siamese twins.
b. inquisitors.
c. pigs.
d. creatures Moreau created.
e. tyrants and criminals.

295. From the passage, it can be inferred that Dr. Moreau
a. does not inflict pain upon animals when he experiments on
them.
b. has caused great pain to the creatures he has experimented on.
c. is unable to experience physical pain.
d. is searching for a way to eliminate physical pain.
e. has learned to feel what an animal feels.

296. Based on the information in the passages, Dr. Moreau is like Victor
Frankenstein in that he also
a. used dead bodies in his experiments.
b. wanted his creations to worship him.
c. made remarkable discoveries.
d. kept his experiment a secret from everyone.
e. had a specific justification for his pursuit of knowledge.

297. Frankenstein would be most upset by Dr. Moreau's
   a. indifference to suffering.
   b. arrogance.
   c. great achievements.
   d. education of animals.
   e. choice of the human form.

298. Which of the following best expresses Frankenstein's and Moreau's attitudes toward science?
   a. Both believe science can be dangerous.
   b. Frankenstein believes science should have a tangible application; Moreau believes scientific knowledge should be sought for its own sake.
   c. Frankenstein believes scientists should not harm living creatures in an experiment; Moreau believes it is acceptable to inflict pain on other creatures.
   d. Both men believe scientists should justify their work.
   e. Both men believe the greatest discoveries often take place in secrecy.

Answers

233. b. The *we* go to school, so the reference must be to school-aged children. In addition, the passage contrasts the *we's* with the *respectable boys* and the *rich ones* (lines 2–3), so the *we's* are neither wealthy nor respected.

234. a. The author and his classmates *go to school through lanes and back streets* (line 1) to avoid the students who go to school dressed in warm and *respectable* clothing. He also states in lines 15–16 that they are *ashamed of the way we look*, implying that they are poorly dressed.

235. d. The boys would get into fights if the rich boys were to utter derogatory words or *pass remarks*.

236. c. While the quote here does show how the author's school masters talked, it has a more important function: to show that his school masters reinforced the class system by telling the author and his classmates to stay in their place and not challenge the existing class structure.

237. e. The author "knows," based only on the fact of which school the boys attend, what they will be when they grow up—the respectable boys will have the administrative jobs (lines 5–6) while the rich boys will *run the government, run the world* (lines 11–12). The author and those in his socio-economic class will be laborers (lines 12–14). The author emphasizes the certainty of this knowledge with the repetition of the phrase *we know* and the sentence *We know that* (line 15). Thus he demonstrates that their future was already set based upon their socio-economic standing.

238. **a.** Lines 6-7 reveal that there are two rooms and lines 9–10 describe the truck *delivering furniture downstairs.*

239. **b.** Lines 1–5 state that after Pauline became pregnant, Cholly had acted like the early days of their marriage when he would ask *if she were tired or wanted him to bring her something from the store.* This statement suggests that Cholly had not done that for a while, and therefore had begun to neglect Pauline.

240. **e** Although there is a *state of ease* (line 5) in the relationship between Pauline and Cholly, there is intense loneliness for Pauline. There may be less tension in this state of ease, but there does not appear to be more intimacy, because the loneliness prevails. We can infer that *back home* she was living with her family, not Cholly, and that Pauline would expect her husband to fulfill her need for companionship.

241. **a.** At the end of the passage, Pauline rediscovers her dreams of romance. Line 14 tells us she *succumbed to her earlier dreams,* and the following sentence tells us what whose dreams were about: *romantic love.*

242. **c.** Because the narrator states that romantic love and physical beauty are *probably the most destructive ideas in the history of human thought* (lines 15–16) because they both *originated in envy, thrived in insecurity, and ended in disillusion,* and because these are the two ideas Pauline was introduced to in the theater, we can infer that she will only become more unhappy as a result of going to the movies.

243. **e.** Lines 4–5 refer to the *reservation jukebox,* and line 12 refers to *the reservation* as well. If Thomas, Chess, and Checkers live on a reservation, they are most likely Native American.

244. **c.** Because their song is one of mourning, **c** is the most logical choice. In addition, the context clue *Samuel was still alive, but* tells us that the song is traditionally reserved for the dead.

245. **c.** To sing a mourning song for someone who is still alive suggests that that person's *life* is mournful—full of grief, sadness, or sorrow.

246. **b.** In line 9, the narrator states that Thomas *wanted his tears to be individual, not tribal,* suggesting too that he felt his father deserved to be mourned as an individual.

247. **e.** The author is speaking figuratively here—the BIA does not literally collect and ferment Indian tears and return them to the reservation in beer and Pepsi cans.

248. **c.** In line 23, the narrator states that Thomas *wanted the songs, the stories, to save everybody*. The paragraph tells readers how many songs Thomas knew but how something seemed to be missing (e.g., he *never sang them correctly*); how Thomas wanted to play the guitar but how *his guitar only sounded like a guitar* (lines 22–23). He wanted his songs to do more, to rescue others.

249. **d.** In lines 15–17, Doc Burton emphasizes change. He tells Mac that *nothing stops* and that as soon as an idea (such as the cause) is put into effect, *it [the idea] would start changing right away*. Then he specifically states that once a commune is established, the *same gradual flux will continue*. Thus, the cause itself is in flux and is always changing.

250. **b.** The several references to communes suggest that the cause is communism, and this is made clear in line 31, when Mac says *Revolution and communism will cure social injustice*.

251. **a.** In lines 21–25, Doc Burton describes his desire to *see the whole picture*, to *look at the whole thing*. He tells Mac he doesn't want to judge the cause as *good* or *bad* so that he doesn't limit his vision. Thus, he is best described as an objective observer.

252. **d.** In the first part of his analogy, Doc Burton says that infections are a reaction to a wound—*the wound is the first battleground* (line 40). Without a wound, there is no place for the infection to fester. The strikes, then, are like the infection in that they are a reaction to a wound (social injustice).

253. **a.** By comparing an individual in a group to a cell within the body (line 50), Doc Burton emphasizes the idea that the individual is really not an individual at all but rather part of a whole.

254. **c.** In lines 59–62, Doc Burton argues that the group *doesn't care* about the *standard* or cause it has created because *the group simply wants to move, to fight*. Individuals such as Mac, however, believe in a cause (or at least think they do).

255. **a.** Doc Burton seems to feel quite strongly that group-man *simply wants to move, to fight*, without needing a real cause—in fact, he states that the group uses the cause *simply to reassure the brains of individual men* (lines 61–62).

256. **b.** Doc Burton knows how deeply Mac believes in the cause and knows that if he outright says *the group doesn't really believe in the cause* that Mac would not listen. Thus he says "*It might be like this*," emphasizing the possibility. Still Mac reacts hotly.

257. **b.** In lines 3–4, Wharton makes it clear that she will be refuting the statement in the first two lines: *but it is certainly a misleading [premise] on which to build any general theory.* In lines 8–9, she states that a subject is suited to a short story *or* a novel, and in lines 9–10, *if it appears to be adapted to both the chances are that it is inadequate in either.* This firmly refutes the opening statement.

258. **d.** After making it clear that subjects are not equally suitable for short stories and novels, Wharton explains what makes a particular subject suitable for the novel form (paragraphs 4 and 5) and how the elements of time and length are different in the short story (paragraph 6).

259. **b.** In lines 15–18, Wharton writes that *rules in art are useful* mainly *for the sake of the guidance they give, but it is a mistake [. . .] to be too much in awe of them.* Thus, they should be used only as a general guide.

260. **a.** Wharton compares *general rules in art* to both *a lamp in a mine* and *a handrail down a black stairwell.*

261. **c.** In paragraph 4, Wharton states the two chief reasons a subject *should find expression in novel-form*: first, *the gradual unfolding of the inner life of its characters* and second *the need of producing in the reader's mind the sense of the lapse of time* (lines 25–27).

262. **b.** Wharton uses this paragraph to clarify the "rules" she established in the previous paragraph by describing more specifically that if a subject can be dealt with in *a single retrospective flash* it is suitable for a short story while those that *justify elaboration* or need to suggest *the lapse of time* require the novel form.

263. **e.** In lines 39–42, Wharton writes that short stories observe *two 'unities'*: that of time, which is limited to achieve *the effect of compactness and instantaneity*, and that of point of view, telling the story *through only one pair of eyes.*

264. **b.** This paragraph expands on the final idea of the previous paragraph, that of the limited point of view. In line 44, Wharton refers to the *character who serves as reflector*—thus in line 46, *this reflecting mind* is that same person, the one who tells the story.

265. **d.** As the introduction states, Higgins is a professor, and he contrasts the life of the gutter with *Science and Literature and Classical Music and Philosophy and Art* (lines 9–10). Thus, his life is best described as the life of a scholar.

266. **e.** The answer to this question is found in Liza's statement in lines 22–24: *You think I must go back to Wimpole Street because I have nowhere else to go but father's.* This statement indicates that Wimpole Street is probably where Liza grew up.

267. **e.** Liza's reply to Higgins suggests that she wants more respect. She criticizes him for always turning everything against her, bullying her, and insulting her. She tells him not to *be too sure that you have me under your feet to be trampled on and talked down* (lines 24–25). Clearly he does not treat her with respect, and as her actions in the rest of the excerpt reveal, she is determined to get it.

268. **b.** Liza is from the *gutter*, but she can't go back there after being with Higgins and living the life of the scholar, a refined, educated, upper-class life. Thus the best definition of *common* here is unrefined.

269. **a.** In these lines Higgins threatens Liza and *lays hands on her*, thus proving that he is a bully.

270. **c.** Higgins refers to Liza as *my masterpiece*, indicating that he thinks of Liza as his creation—that he made her what she is today.

271. **b.** The excerpt opens with Higgins telling Liza "*If you're going to be a lady*" and comparing her past—the *life of the gutter*—with her present—a cultured life of literature and art. We also know that Higgins taught Liza phonetics (line 40) and that Liza was once *only a flower girl* but is now a duchess (lines 55–56). Thus, we can conclude that Higgins taught Liza how to speak and act like someone from the upper class.

272. **d.** Higgins realizes that Liza—with the knowledge that he gave her—now has the power to stand up to him, that she will not just let herself be *trampled on and called names* (line 59). He realizes that she has other options and she is indifferent to his *bullying and big talk* (line 55).

273. **c.** Liza's final lines express her joy at realizing that she has the power to change her situation and that she is not Higgins' inferior but his equal; she can't believe that *all the time I had only to lift up my finger to be as good as you* (lines 59–60). She realizes that she can be an assistant to someone else, that she doesn't have to be dependent on Higgins.

274. **d.** In the first few lines, the narrator states that Miss Temple was the *superintendent of the seminary* and that she received both *instruction* and *friendship* from Miss Temple, who was also like a mother to her *she had stood me in the stead of mother*.

275. **a.** The narrator states that with Miss Temple, *I had given in allegiance to duty and order; I was quiet; I believed I was content* (lines 12–13).

276. **d.** The context here suggests existence or habitation, not captivity or illness.

277. **c.** We can assume that the narrator would go home during vacations, but she spent all of her vacations at school because *Mrs. Reed had* *never sent for me to Gateshead* (lines 50–51). Thus we can infer that Mrs. Reed was her guardian, the one who sent the narrator to Lowood in the first place.

278. b. The narrator describes her experience with *school-rules* and *school-duties* (line 53) and how she *tired of the routine* (line 56) after Miss Temple left. She also contrasts Lowood with the *real world* of *hopes and fears, of sensations and excitements* (lines 35–36) and that the view from her window seemed a *prison-ground, exile limits* (line 44). Thus, it can be inferred that Lowood is both a structured and isolated place.

279. a. The narrator states in lines 26–27 that she had *undergone a transforming process* and that now she again felt *the stirring of old emotions* (line 30) and *remembered that the real world was wide* and *awaited those who had courage to go forth* (lines 36–37). She also looks at the road from Lowood and states *how [she] longed to follow it further!* More importantly, she repeats her desire for *liberty* and prays for *a new servitude*—something beyond Lowood.

280. e. In lines 13–15, the narrator states that with Miss Temple at Lowood, she *believed* she was content, that *to the eyes of others, usually even to my own, I appeared a disciplined and subdued character*. This suggests that in her *natural element* (lines 29–30) she is not so disciplined or subdued. Her desire for freedom and to explore the world are also evident in this passage; she longs to follow the road that leads away from Lowood (line 46) and she is *half desperate* in her cry for something new, something beyond Lowood and the rules and systems she *tired of [. . .] in one afternoon* (line 56).

281. d. Because Lowood had been the narrator's home for eight years and all she *knew of existence* was school rules, duties, habits, faces, etc. (lines 53–55)—because she had had *no communication [. . .] with the outer world* (lines 52–53), it is likely that she feels her initial prayers were unrealistic. At least a *new servitude* would provide some familiar territory, and it therefore seems more realistic and attainable than *liberty* or *change*.

282. c. The women refer to each other as "Mrs.", and their conversation reveals that they don't know much about each other. Mrs. Hale, for example, asks Mrs. Peters if she knew Mr. Wright line 46) and if she were *raised round here* (line 58).

283. a. Mrs. Peters says *It would be lonesome for me sitting here alone* (lines 27–28)—to which Mrs. Hale replies, *It would, wouldn't it?* and then expresses her wish that she'd come to see Mrs. Wright. She says *it's a lonesome place and always was* in line 37 and then says *I can see now*—(lines 38–39) suggesting that she can understand now how Mrs. Wright must have felt.

284. **d.** Mrs. Hale describes Mr. Wright as a *hard man* who was *like a raw wind that gets to the bone* (lines 51–52). Mrs. Wright's loneliness would be deepened by living with a man who was quiet and cold.

285. **b.** The punctuation here—the dashes between each word—suggest that Mrs. Wright changed from the sweet, fluttery woman she was to a bitter, unhappy person over the years. The emphasis on her loneliness and the dead husband and bird add to this impression.

286. **d.** The women decide to take the quilt to Mrs. Wright to keep her busy; it would give her something to do, something familiar and comforting

287. **c.** Because her house was so lonely, Mrs. Wright would have wanted the company of a pet—and a pet that shared some qualities with her (or with her younger self) would have been particularly appealing. She would have liked the bird's singing to ease the quiet in the house, and she also *used to sing real pretty herself* (line 10) and would have felt a real connection with the bird.

288. **b.** The clues in the passage—the violently broken bird cage, the dead bird lovingly wrapped in silk and put in a pretty box, the description of John Wright as a hard and cold man—suggest that he killed the bird and that Mrs. Wright in turn killed him for destroying her companion.

289. **d.** The fact that Mrs. Hale *slips box under quilt pieces* suggests that she will not share her discovery with the men.

290. **c.** Frankenstein asks his listener to *[l]earn from me [. . .] how dangerous is the acquirement of knowledge* (lines 6–8). He is telling his tale as a warning and does not want to lead his listener into the same kind of *destruction and infallible misery* (line 6).

291. **a.** The context reveals that Frankenstein was prepared for *a multitude of reverses* or setbacks that would hinder his operations.

292. **e.** Frankenstein describes himself as pursuing his *undertaking with unremitting ardour* and that his *cheek had grown pale with study, and [his] person had become emaciated with confinement* (lines 45–47). He also says that a *resistless, and almost frantic, impulse urged me forward; I seemed to have lost all soul or sensation but for this one pursuit* (lines 56–58). These are the marks of a man obsessed.

293. **b.** Moreau states in lines 22–24 that *this extraordinary branch of knowledge has never been sought as an end, [. . .] until I took it up!*, and in lines 28–30, he states that he was *the first man to take up this question armed with antiseptic surgery, and with a really scientific knowledge* *of the laws of growth.* This, and the detail with which he explains the background of his investigations, reveal that he is a calculating and systematic scientist. (Although he *confesses* that he chose the human form *by chance* (line 45), it is likely that Moreau *did not* just happen upon this choice but that he found the human form, as he later states, more appealing *to the artistic turn of mind [. . .] than any animal shape* (lines 48–49).

294. **d.** Right after he says *these things*, the narrator says *these animals* to clarify that he is referring to the creatures that Moreau created. An additional context clue is provided by Moreau's response, in which he explains how animals may be *educated* so that they may talk.

295. **b.** The narrator asks Moreau to justify *all this pain* (line 54), implying that he has inflicted great pain on the animals he has used in his experiments.

296. **c.** Both men make remarkable discoveries in their fields; in the other aspects the men are different. Dr. Moreau uses live animals to change their form, and there is no evidence in the passage that he wants his creatures to worship him or that he has kept his experiment a secret (though these facts *are* evident in other passages in the book). Passage 2 also suggests that Moreau did not have a specific application or justification for his work; he responds to the narrator's request for a justification by philosophizing about pain.

297. **a.** Frankenstein confesses that he was horrified by the torture of living animals that that he trembled just remembering the pain he inflicted (lines 52–55). He also characterizes himself as having *lost all soul or sensation* (line 57) in his quest. In addition, he is telling this tale as a warning. Thus it is likely that he would be most offended by Moreau's indifference to the suffering of other creatures.

298. **b.** In lines 29–35, Frankenstein cites specific goals for his pursuit of knowledge: he wanted to *pour a torrent of light into our dark world* by making important new discoveries; he wanted to create a new species that would *bless [him] as its creator and source*; and he wanted to *renew life.* Moreau, on the other hand, does not offer any application or justification; he seems motivated only by the acquisition of knowledge. He states that he has *devoted* his life to *the study of the plasticity of living forms* (lines 2–3) and seems more interested in *what science has to teach* (lines 65–66) than in what can be done with that knowledge. This is reinforced by the fact that he does not offer a justification for his experiments.

Music

Questions 299–303 are based on the following passage.

The following passage describes the transition from the swing era to bebop in the history of jazz music.

(1)   Jazz, from its early roots in slave spirituals and the marching bands of New Orleans, had developed into the predominant American musical style by the 1930s. In this era, jazz musicians played a lush, orchestrated style known as swing. Played in large ensembles, also called big bands,

(5)   swing filled the dance halls and nightclubs. Jazz, once considered risqué, was made more accessible to the masses with the vibrant, swinging sounds of these big bands. Then came bebop. In the mid-1940s, jazz musicians strayed from the swing style and developed a more improvisational method of playing known as bebop. Jazz was transformed from

(10)   popular music to an elite art form.

     The soloists in the big bands improvised from the melody. The young musicians who ushered in bebop, notably trumpeter Dizzy Gillespie and saxophonist Charlie Parker, expanded on the improvisational elements of the big bands. They played with advanced harmonies,

(15)   changed chord structures, and made chord substitutions. These young musicians got their starts with the leading big bands of the day, but during World War II—as older musicians were drafted and dance halls made cutbacks—they started to play together in smaller groups.

(20) These pared-down bands helped foster the bebop style. Rhythm is the distinguishing feature of bebop, and in small groups the drums became more prominent. Setting a driving beat, the drummer interacted with the bass, piano, and the soloists, and together the musicians created fast, complex melodies. Jazz aficionados flocked to such clubs as Minton's Playhouse in Harlem to soak in the new style. For the young

(25) musicians and their fans this was a thrilling turning point in jazz history. However, for the majority of Americans, who just wanted some swinging music to dance to, the advent of bebop was the end of jazz as mainstream music.

299. The swing style can be most accurately characterized as
  a. complex and inaccessible.
  b. appealing to an elite audience.
  c. lively and melodic.
  d. lacking in improvisation.
  e. played in small groups.

300. According to the passage, in the 1940s you would most likely find bebop being played where?
  a. church
  b. a large concert hall
  c. in music schools
  d. small clubs
  e. parades

301. According to the passage, one of the most significant innovations of the bebop musicians was
  a. to shun older musicians.
  b. to emphasize rhythm.
  c. to use melodic improvisation.
  d. to play in small clubs.
  e. to ban dancing.

302. In the context of this passage, *aficionados* (line 23) can most accurately be described as
  a. fans of bebop.
  b. residents of Harlem.
  c. innovative musicians.
  d. awkward dancers.
  e. fickle audience members.

303. The main purpose of the passage is to
  a. mourn the passing of an era.
  b. condemn bebop for making jazz inaccessible.
  c. explain the development of the bebop style.
  d. celebrate the end of the conventional swing style of jazz.
  e. instruct in the method of playing bebop.

Questions 304–309 are based on the following passage.

This passage details the rise and fall of the Seattle grunge-music sound in American pop culture of the 1990s.

(1)   The late 1980s found the landscape of popular music in America dominated by a distinctive style of rock and roll known as *Glam Rock* or *Hair Metal*—so called because of the over-styled hair, makeup, and wardrobe worn by the genre's ostentatious rockers. Bands like Poison,
(5)   White Snake, and Mötley Crüe popularized glam rock with their power ballads and flashy style, but the product had worn thin by the early 1990s. The mainstream public, tired of an act they perceived as symbolic of the superficial 1980s, was ready for something with a bit of substance.
(10)   In 1991, a Seattle-based band named Nirvana shocked the corporate music industry with the release of its debut single, "Smells Like Teen Spirit," which quickly became a huge hit all over the world. Nirvana's distorted, guitar-laden sound and thought-provoking lyrics were the antithesis of glam rock, and the youth of America were quick to pledge
(15)   their allegiance to the brand new movement known as *grunge*.
        Grunge actually got its start in the Pacific Northwest during the mid 1980s, the offspring of the metal-guitar driven rock of the 1970s and the hardcore, punk music of the early 1980s. Nirvana had simply brought into the mainstream a sound and culture that got its start
(20)   years before with bands like Mudhoney, Soundgarden, and Green River. Grunge rockers derived their fashion sense from the youth culture of the Pacific Northwest: a melding of punk rock style and outdoors clothing like flannels, heavy boots, worn out jeans, and corduroys. At the height of the movement's popularity, when other
(25)   Seattle bands like Pearl Jam and Alice in Chains were all the rage, the trappings of grunge were working their way to the height of American fashion. Like the music, teenagers were fast to embrace the grunge fashion because it represented defiance against corporate America and shallow pop culture.

(30)   Many assume that grunge got its name from the unkempt appearance of its musicians and their dirty, often distorted guitar sounds. However, rock writers and critics have used the word "grunge" since the 1970s. While no one can say for sure who was the first to characterize a Seattle band as "grunge," the most popular theory is that it
(35)   originated with the lead singer of Mudhoney, Mark Arm. In a practical joke against a local music magazine, he placed advertisements all over Seattle for a band that did not exist. He then wrote a letter to the magazine complaining about the quality of the fake band's music. The magazine published his critique, one part of which stated, "I hate Mr.
(40)   Epp and the Calculations! Pure grunge!"

   The popularity of grunge music was ephemeral; by the mid- to late-1990s its influence upon American culture had all but disappeared, and most of its recognizable bands were nowhere to be seen on the charts. The heavy sound and themes of grunge were replaced on the radio
(45)   waves by bands like NSYNC, the Backstreet Boys, and the bubblegum pop of Britney Spears and Christina Aguilera.

   There are many reasons why the Seattle sound faded out of the mainstream as quickly as it rocketed to prominence, but the most glaring reason lies at the defiant, anti-establishment heart of the
(50)   grunge movement itself. It is very hard to buck the trend when you are the one setting it, and many of the grunge bands were never comfortable with the celebrity that was thrust upon them. One the most successful Seattle groups of the 1990s, Pearl Jam, filmed only one music video, and refused to play large venues. Ultimately, the simple
(55)   fact that many grunge bands were so against mainstream rock stardom eventually took the movement back to where it started: underground. The American mainstream public, as quick as they were to hop onto the grunge bandwagon, were just as quick to hop off, and move onto something else.

304.   The author's description of glam rockers (lines 2–7) indicates that they
   a. cared more about the quality of their music than money.
   b. were mainly style over substance.
   c. were unassuming and humble.
   d. were songwriters first, and performers second.
   e. were innovators in rock and roll.

305. The word *ostentatious* in line 4 most nearly means
   a. stubborn.
   b. youthful.
   c. showy.
   d. unadorned.
   e. popular.

306. In lines 25–26, the phrase *the trappings of grunge* refers to
   a. the distorted sound of grunge music.
   b. what the grunge movement symbolized.
   c. the unattractiveness of grunge fashion.
   d. the clothing typical of the grunge movement.
   e. the popularity of grunge music.

307. Which of the following is not associated with the grunge movement?
   a. Mr. Epps and the Calculations
   b. Pearl Jam
   c. Nirvana
   d. Green River
   e. White Snake

308. Which of the following words best describes the relationship between grunge music and its mainstream popularity?
   a. solid
   b. contrary
   c. enduring
   d. acquiescent
   e. unprofitable

309. In line 41, the word *ephemeral* most nearly means
   a. enduring.
   b. unbelievable.
   c. a fluke.
   d. fleeting.
   e. improbable.

Questions 310–316 are based on the following passage.

The selection that follows is based on an excerpt from the biography of a music legend.

(1)   Although Dick Dale is best known for his contributions to surf music, and has been called "King of the Surf Guitar," he has also been referred to as the "Father of Heavy Metal." While this title is more often associated with Ozzy Osbourne or Tony Iossa, Dale earned it from *Guitar*
(5)   *Player Magazine* for his unique playing style and pioneering use of Fender guitars and amplifiers.

          In the mid-1950s, Dale was playing guitar at a club in California, where his one-of-a-kind music turned it from a jazz club into a rock nightspot. After a 1956 concert there, guitar and amplifier maker Leo Fender
(10)   approached the guitarist and gave him the first Fender Stratocaster to try before the guitar was mass marketed. Fender thought that Dale's way of playing, a virtual assault on the instrument, would be a good test of its durability. However, the guitar was right-handed and Dale played left-handed. Unfazed, Dale held and played it upside down and backwards (a
(15)   feat that later strongly influenced Jimi Hendrix).

          The test proved too much for Fender's equipment. Dale loved the guitar, but blew out the amplifier that came with it. It had worked well for most other musicians, who at that time were playing country and blues. Rock didn't exist, and no one played the guitar as fiercely as Dale. Fender
(20)   improved the amplifier, and Dale blew it out again. Before Fender came up with a winner, legend has it that Dale blew up between 40 and 60 amplifiers. Finally, Fender created a special amp just for Dale, known as the "Showman." It had more than 100 watts of power. The two men then made an agreement that Dale would "road test" prototypes of Fender's
(25)   new amplification equipment before they would be manufactured for the general public. But they still had problems with the speakers—every speaker Dale used it with blew up (some even caught fire) because of the intense power of his volume coupled with a staccato playing style.

          Fender and Dale approached the James B. Lansing speaker company,
(30)   asking for a fifteen-inch speaker built to their specifications. The company responded with the fifteen-inch JBL-D130F speaker, and it worked. Dale was able to play through the Showman Amp with the volume turned all the way up. With the help of Leo Fender and the designers at Lansing, Dick Dale was able to break through the limits of existing
(35)   electronics and play the music his way—loud.

          But it wasn't enough. As Dale's popularity increased, his shows got larger. He wanted even more sound to fill the larger halls he now played in. Fender had the Triad Company craft an amp tube that

(40) peaked at 180 watts, creating another new amplifier for Dale. Dale designed a cabinet to house it along with two Lansing speakers. He called it the Dick Dale Transformer, and it was a scream machine. Dick Dale made music history by playing a new kind of music, and helping to invent the means by which that music could be played. Not only was this the start of the electric movement, but it may also be considered
(45) the dawning of heavy metal.

310. In line 25, the word *prototype* most nearly means
 a. an original model.
 b. a Fender guitar.
 c. an amplifier-speaker combination.
 d. a computerized amplifier.
 e. top of the line equipment.

311. Lines 16–20 indicate that
 a. country and blues guitarists didn't need amplifiers.
 b. most musicians played louder than Dick Dale.
 c. a new kind of music was being created.
 d. Dick Dale needed a new guitar.
 e. the Stratocaster didn't work for Dick Dale.

312. In line 28, the word *staccato* most nearly means
 a. smooth and connected.
 b. loud.
 c. gently picking the guitar strings.
 d. abrupt and disconnected.
 e. peaceful.

313. The title that best suits this passage is
 a. Dick Dale and the History of the Amplifier.
 b. The King of Heavy Metal.
 c. The Invention of the Stratocaster.
 d. Lansing and Fender: Making Music History.
 e. How Surf Music Got its Start.

314. In line 14, *unfazed* most nearly means
 a. not moving forward.
 b. not in sequence.
 c. not bothered by.
 d. not ready for.
 e. not happy about.

315. In line 41, *scream machine* indicates that
 a. the new transformer could handle very loud music.
 b. fans screamed when they heard Dale play.
 c. Dale's guitar sounded like it was screaming.
 d. neighbors of the club screamed because the music was too loud.
 e. you couldn't hear individual notes being played.

316. All of the following can explicitly be answered on the basis of the passage EXCEPT
 a. Who invented the Stratocaster?
 b. Where did Dick Dale meet Leo Fender?
 c. What company made speakers for Dick Dale?
 d. Where did Ozzy Osbourne get his start as a musician?
 e. What do Dick Dale, Ozzy Osbourne, and Tony Iossa have in common?

Questions 317–323 are based on the following passage.

The following passage discusses the unique musical traditions that developed along the Rio Grand in colonial New Mexico.

(1)   From 1598 to 1821, the area along the Rio Grand that is now the state of New Mexico formed the northernmost border of the Spanish colonies in the New World. The colonists lived on a geographic frontier surrounded by deserts and mountains. This remote colony with its
(5)   harsh climate was far removed from the cultural centers of the Spanish Empire in the New World, and music was a necessary part of social life. The isolated nature of the region and needs of the community gave rise to a unique, rich musical tradition that included colorful ballads, popular dances, and some of the most extraordinary ceremonial
(10)   music in the Hispanic world.
       The popular music along the Rio Grand, especially the heroic and romantic ballads, reflected the stark and rough nature of the region. Unlike the refined music found in Mexico, the music of the Rio Grand had a rough-cut "frontier" quality. The music also reflected the mix-
(15)   ing of cultures that characterized the border colony. The close military and cultural ties between the Spanish and the native Pueblos of the region led to a uniquely New Mexican fusion of traditions. Much of the music borrowed from both European and native cultures. This mixing of traditions was especially evident in the dances.
(20)   The *bailes*, or village dances—instrumental music played on violin and guitar—were a lively focus of frontier life. Some *bailes* were derived from traditional European waltzes, but then adapted to the

singular style of the region. The *bailes* had an unusual melodic structure and the players had unique methods of bowing and tuning their
(25) instruments. Other *bailes*, such as *indita* (little Indian girl) and *vaquero*
(cowboy), were only found in New Mexico. The rhythms and
melodies of the *indita* had definite Puebloan influences. Its themes,
which ranged from love to tragedy, almost always featured dramatic
interactions between Spanish and Native Americans. Similarly, the
(30) *Matachines* dance drama was an allegorical representation of the meeting of European and Native American cultures. Its European
melodies, played on violin and guitar, were coupled with the use of
insistent repetition, which came from the Native American tradition.

In addition to the *bailes*, waltzes—the Waltz of the Days and the
(35) Waltz of the Immanuels—were also performed to celebrate New
Year's Eve and New Year's Day. Groups of revelers went singing from
house to house throughout the night to bring in the New Year. In New
Mexico, January 1 is the Feast of Immanuel so the singers visited the
houses of people named Manuel or Manuela. Many songs were sung
(40) on these visits but especially popular were the *coplas*, or improvised
couplets, composed on the spot to honor or poke fun of the person
being visited.

Like in the New Year's celebration, music was central to many social
rituals in colonial New Mexico. In the Rio Grand region, weddings
(45) were performed in song in a folk ceremony called "The Delivery of
the Newlyweds." The community would gather to sanction the new
couple and "deliver" them in song to each other and to their respective families. The verses of the song, played to a lively waltz, were
improvised, but followed a familiar pattern. The first verses spoke
(50) about marriage in general. These were followed by serious and
humorous verses offering practical advice to the couple. Then all the
guests filed past to bless the couple and concluding verses were sung
to honor specific individuals such as the best man. At the wedding
dance, *la marcha* was performed. In this triumphal march, couples
(55) formed into single files of men and women. After dancing in concen-
tric circles, the men and women lined up opposite one another with
their hands joined overhead to form a tunnel of love from which the
new couple was the last to emerge.

By the turn of the twentieth century, styles were evolving and musi-
(60) cal forms popular in previous eras were giving way to new tastes. The
ancient romance ballads were replaced by newer forms that featured
more local and contemporary events. The extraordinary *indita* was no
longer performed and the *canción*, or popular song, had begun its rise.
However, many of the wedding traditions of the colonial era are still

(65)   in practice today. The music that was so central to life in the remote colony of New Mexico has much to teach us about the unique and vibrant culture that once flourished there.

317. The primary purpose of the first paragraph is to
   a. describe the geography of New Mexico.
   b. instruct readers about the history of the Spanish colonies along the Rio Grand.
   c. introduce readers to the unique culture and musical traditions along the Rio Grand.
   d. list the types of music that were prevalent in colonial New Mexico.
   e. explain the unique musical traditions of the New Mexican colonies.

318. In line 23, the word *singular* most nearly means
   a. strange.
   b. monotone.
   c. separate.
   d. unusual.
   e. superior.

319. According to the passage, the musical tradition found in New Mexico was the result of all the following EXCEPT
   a. distance from cultural centers.
   b. the blending of cultures.
   c. the geography of the region.
   d. the imposition of European culture on native traditions.
   e. unique ways of playing instruments.

320. The New Year's celebration and wedding ceremony described in the passage share in common
   a. offering of practical advice.
   b. use of a lively march.
   c. use of improvised verses.
   d. visiting of houses.
   e. singing and dancing.

321. According to the passage, the main purpose of the "Delivery of the Newlyweds" was to
a. sanction and bless the new couple.
b. form a tunnel of love.
c. marry couples who did not want a Church wedding.
d. offer advice to the new couple.
e. sing improvised songs to newlyweds.

322. Which of the titles provided below is most appropriate for this passage?
a. Wedding Marches and New Year's Waltzes of the Rio Grand
b. The Fading Era of Colonial Music in New Mexico
c. Cowboy Songs of the Past
d. Between Deserts and Mountains New Mexico Sings a Unique Song
e. The Extraordinary Popular and Ceremonial Music of the Rio Grand

323. The author's attitude toward the music of colonial New Mexico can best be described as
a. bemusement.
b. admiration.
c. alienation.
d. condescension.
e. awe.

Questions 324–332 are based on the following passages.

In Passage 1, the author describes the life and influence of blues guitarist Robert Johnson. In Passage 2, the author provides a brief history of the blues.

PASSAGE 1

(1)   There is little information available about the legendary blues guitarist Robert Johnson, and the information that *is* available is as much rumor as fact. What is undisputable, however, is Johnson's impact on the world of rock and roll. Some consider Johnson the father of modern
(5)   rock; his influence extends to artists from Muddy Waters to Led Zeppelin, from the Rolling Stones to the Allman Brothers Band. Eric Clapton, arguably the greatest living rock guitarist, has said that "Robert Johnson to me is the most important blues musician who ever

(10) lived. [. . .] I have never found anything more deeply soulful than Robert Johnson." While the impact of Johnson's music is evident, the genesis of his remarkable talent remains shrouded in mystery.

For Johnson, born in 1911 in Hazelhurst, Mississippi, music was a means of escape from working in the cotton fields. As a boy he worked on the farm that belonged to Noel Johnson—the man rumored to be

(15) his father. He married young, at age 17, and lost his wife a year later in childbirth. That's when Johnson began traveling and playing the blues.

Initially Johnson played the harmonica. Later, he began playing the guitar, but apparently he was not very good. He wanted to learn, how-

(20) ever, so he spent his time in blues bars watching the local blues legends Son House and Willie Brown. During their breaks, Johnson would go up on stage and play. House reportedly thought Johnson was so bad that he repeatedly told Johnson to get lost. Finally, one day, he did. For six months, Johnson mysteriously disappeared. No one knew what

(25) happened to him.

When Johnson returned half a year later, he was suddenly a first-rate guitarist. He began drawing crowds everywhere he played. Johnson never revealed where he had been and what he had done in those six months that he was gone. People had difficulty understanding how

(30) he had become so good in such a short time. Was it genius? Magic? Soon, rumors began circulating that he had made a deal with the devil. Legend has it that Johnson met the devil at midnight at a crossroads and sold his soul to the devil so he could play guitar.

Johnson recorded only 29 songs before his death in 1938, purport-

(35) edly at the hands of a jealous husband. He was only 27 years old, yet he left an indelible mark on the music world. There are countless versions of "Walkin' Blues," and his song "Cross Road Blues" (later retitled "Crossroads") has been recorded by dozens of artists, with Cream's 1969 version of "Crossroads" being perhaps the best-known

(40) Johnson remake. Again and again, contemporary artists return to Johnson, whose songs capture the very essence of the blues, transforming our pain and suffering with the healing magic of his guitar.

PASSAGE 2

(1) There are more than fifty types of blues music, from the famous Chicago and Memphis Blues to the less familiar Juke Joint and Acoustic Country Blues. This rich variety comes as no surprise to those who recognize the blues as a fundamental American art form.

(5) Indeed, in its resolution to name 2003 the Year of the Blues, the 107th

Congress has declared that the blues is "the most influential form of American roots music." In fact, the two most popular American musical forms—rock and roll and jazz—owe their genesis in large part (some would argue entirely) to the blues.

(10) The blues—a neologism attributed to the American writer Washington Irving (author of "The Legend of Sleepy Hollow") in 1807—evolved from black American folk music. Its beginnings can be traced to songs sung in the fields and around slave quarters on southern plantations, songs of pain and suffering, of injustice, of longing for a bet-
(15) ter life. A fundamental principle of the blues, however, is that the music be cathartic. Listening to the blues will drive the blues away; it is music that has the power to overcome sadness. Thus "the blues" is something of a misnomer, for the music is moving but not melancholy; it is, in fact, music born of hope, not despair.

(20) The blues began to take shape as a musical movement in the years after emancipation, around the turn of the century when blacks were technically free but still suffered from social and economic discrimination. Its poetic and musical forms were popularized by W. C. Handy just after the turn of the century. Handy, a classical guitarist who
(25) reportedly heard the blues for the first time in a Mississippi train station, was the first to officially compose and distribute "blues" music throughout the United States, although its popularity was chiefly among blacks in the South. The movement coalesced in the late 1920s and indeed became a national craze with records by blues singers such
(30) as Bessie Smith selling in the millions.

The 1930s and 1940s saw a continued growth in the popularity of the blues as many blacks migrated north and the blues and jazz forms continued to develop, diversify, and influence each other. It was at this time that Son House, Willie Brown, and Robert Johnson played, while
(35) the next decade saw the emergence of the blues greats Muddy Waters, Willie Dixon, and Johnny Lee Hooker.

After rock and roll exploded on the music scene in the 1950s, many rock artists began covering blues songs, thus bringing the blues to a young white audience and giving it true national and international
(40) exposure. In the early 1960s, the Rolling Stones, Yardbirds, Cream, and others remade blues songs such as Robert Johnson's "Crossroads" and Big Joe Williams' "Baby Please Don't Go" to wide popularity. People all across America—black and white, young and old, listened to songs with lyrics that were intensely honest and personal, songs that
(45) told about any number of things that give us the blues: loneliness, betrayal, unrequited love, a run of bad luck, being out of work or away from home or broke or broken hearted. It was a music perfectly suited

(50) for a nation on the brink of the Civil Rights movement—a kind of
music that had the power to cross boundaries, to heal wounds, and to
offer hope to a new generation of Americans.

324. In Passage 1, the author's main goal is to
  a. solve the mystery of the genesis of Johnson's talent.
  b. provide a detailed description of Johnson's music and style.
  c. provide a brief overview of Johnson's life and influence.
  d. prove that Johnson should be recognized as the greatest blues
    musician who ever lived.
  e. explain how Johnson's music impacted the world of rock and roll.

325. The information provided in the passage suggests that Johnson
  a. really did make a deal with the devil.
  b. was determined to become a great guitarist, whatever the cost.
  c. wasn't as talented as we have been led to believe.
  d. disappeared because he had a breakdown.
  e. owes his success to Son House and Willie Brown.

326. The word *neologism* in Passage 2, line 10 means
  a. a new word or use of a word.
  b. a grassroots musical form.
  c. a fictional character or fictitious setting.
  d. the origin or source of something.
  c. the evolution of a person, place, or thing.

327. In Passage 2, the sentence *People all across America—black and white,
  young and old, listened to songs with lyrics that were intensely honest and
  personal, songs that told about any number of things that give us the
  blues: loneliness, betrayal, unrequited love, a run of bad luck, being out of
  work or away from home or broke or broken hearted* (lines 43–47), the
  author is
  a. defining blues music.
  b. identifying the origin of the blues.
  c. describing the lyrics of a famous blues song.
  d. explaining why blues remakes were so popular.
  e. making a connection between the blues and the Civil Rights
    movement.

328. In the last paragraph of Passage 2 (lines 37–50), the author suggests that
  a. the blues should be recognized as more important and complex musical form than rock and roll.
  b. the golden age of rock and roll owes much to the popularity of blues cover songs.
  c. music has always been a means for people to deal with intense emotions and difficulties.
  d. a shared interest in the blues may have helped blacks and whites better understand each other and ease racial tensions.
  e. the rock and roll versions of blues songs were better than the originals.

329. Both authors would agree on all of the following points EXCEPT
  a. listening to the blues is cathartic.
  b. Robert Johnson is the best blues guitarist from the 1930s and 1940s.
  c. the blues are an important part of American history.
  d. "Crossroads" is one of the most well-known blues songs.
  e. blues music is deeply emotional.

330. The passages differ in tone and style in that
  a. Passage 1 is intended for a general audience while Passage 2 is intended for readers with a musical background.
  b. Passage 1 is far more argumentative than Passage 2.
  c. Passage 1 is often speculative while Passage 2 is factual and assertive.
  d. Passage 1 is more formal than Passage 2, which is quite casual.
  e. Passage 1 is straight-forward while Passage 2 often digresses from the main point.

331. Which of the following best describes the relationship between these two passages?
  a. specific : general
  b. argument : support
  c. fiction : nonfiction
  d. first : second
  e. cause : effect

332. Which of the following sentences from Passage 2 could most effectively be added to Passage 1?

a. *In fact, the two most popular American musical forms—rock and roll and jazz—owe their genesis in large part (some would argue entirely) to the blues.* (lines 7–9)

b. *A fundamental principle of the blues, however, is that the music be cathartic.* (line 15–16)

c. *Thus "the blues" is something of a misnomer, for the music is moving but not melancholy; it is, in fact, music born of hope, not despair.* (lines 17–19)

d. *It was at this time that Son House, Willie Brown, and Robert Johnson played, while the next decade saw the emergence of the blues greats Muddy Waters, Willie Dixon and Johnny Lee Hooker.* (lines 33–36)

e. *After rock and roll exploded on the music scene in the 1950s, many rock artists began covering blues songs, thus bringing the blues to a young white audience and giving it true national and international exposure.* (lines 37–40)

Questions 333–342 are based on the following passage.

This passage describes the formative experiences of the composer Wolfgang Amadeus Mozart.

(1)   The composer Wolfgang Amadeus Mozart's remarkable musical talent was apparent even before most children can sing a simple nursery rhyme. Wolfgang's older sister Maria Anna, who the family called Nannerl, was learning the clavier, an early keyboard instrument, when *(5)* her three-year-old brother took an interest in playing. As Nannerl later recalled, Wolfgang "often spent much time at the clavier, picking out thirds, which he was always striking, and his pleasure showed that it sounded good." Their father Leopold, an assistant concertmaster at the Salzburg Court, recognized his children's unique gifts *(10)* and soon devoted himself to their musical education.

Born in Salzburg, Austria, on January 27, 1756, Wolfgang was five when he learned his first musical composition—in less than half an hour. He quickly learned other pieces, and by age five composed his first original work. Leopold settled on a plan to take Nannerl and *(15)* Wolfgang on tour to play before the European courts. Their first venture was to nearby Munich where the children played for Maximillian III Joseph, elector of Bavaria. Leopold soon set his sights on the capital of the Hapsburg Empire, Vienna. On their way to Vienna, the

(20) family stopped in Linz, where Wolfgang gave his first public concert. By this time, Wolfgang was not only a virtuoso harpsichord player but he had also mastered the violin. The audience at Linz was stunned by the six-year-old, and word of his genius soon traveled to Vienna. In a much-anticipated concert, the children appeared at the Schönbrunn Palace on October 13, 1762. They utterly charmed the emperor and (25) empress.

Following this success, Leopold was inundated with invitations for the children to play, for a fee. Leopold seized the opportunity and booked as many concerts as possible at courts throughout Europe. After the children performed at the major court in a region, other (30) nobles competed to have the "miracle children of Salzburg" play a private concert in their homes. A concert could last three hours, and the children played at least two a day. Today, Leopold might be considered the worst kind of stage parent, but at the time it was not uncommon for prodigies to make extensive concert tours. Even so, it was an (35) exhausting schedule for a child who was just past the age of needing an afternoon nap.

Wolfgang fell ill on tour, and when the family returned to Salzburg on January 5, 1763, Wolfgang spent his first week at home in bed with acute rheumatoid arthritis. In June, Leopold accepted an invitation for (40) the children to play at Versailles, the lavish palace built by Loius XIV, king of France. Wolfgang did not see his home in Salzburg for another three years. When they weren't performing, the Mozart children were likely to be found bumping along the rutted roads in an unheated carriage. Wolfgang passed the long uncomfortable hours in the imaginary (45) Kingdom of Back, of which he was king. He became so engrossed in the intricacies of his make-believe court that he persuaded a family servant to make a map showing all the cities, villages, and towns over which he reigned.

The king of Back was also busy composing. Wolfgang completed (50) his first symphony at age nine and published his first sonatas that same year. Before the family returned to Salzburg, Wolfgang had played for, and amazed, the heads of the French and British royal families. He had also been plagued with numerous illnesses. Despite Wolfgang and Nannerl's arduous schedule and international renown, the family's (55) finances were often strained. The pattern established in his childhood would be the template for the rest of his short life. Wolfgang Amadeus Mozart toiled constantly, was lauded for his genius, suffered from illness, and struggled financially, until he died at age 35. The remarkable child prodigy who more than fulfilled his potential was buried in an (60) unmarked grave, as was the custom at the time, in a Vienna suburb.

333. The primary purpose of the passage is to
   a. illustrate the early career and formative experiences of a musical prodigy.
   b. describe the classical music scene in the eighteenth century.
   c. uncover the source of Wolfgang Amadeus Mozart's musical genius.
   d. prove the importance of starting a musical instrument at an early age.
   e. denounce Leopold Mozart for exploiting his children's talent.

334. According to the passage, Wolfgang became interested in music because
   a. his father thought it would be profitable.
   b. he had a natural talent.
   c. he saw his sister learning to play.
   d. he came from a musical family.
   e. he wanted to go on tour.

335. What was the consequence of Wolfgang's first public appearance?
   a. He charmed the emperor and empress of Hapsburg.
   b. Leopold set his sights on Vienna.
   c. Word of Wolfgang's genius spread to the capital.
   d. He mastered the violin.
   e. Invitations for the "miracle children" to play poured in.

336. The author's attitude toward Leopold Mozart can best be characterized as
   a. vehement condemnation.
   b. mild disapproval.
   c. glowing admiration.
   d. incredulity.
   e. veiled disgust.

337. In line 40, the word *lavish* most nearly means
   a. wasteful.
   b. clean.
   c. extravagant.
   d. beautiful.
   e. glorious.

338. The author uses the anecdote about Mozart's Kingdom of Back to illustrate
  a. Mozart's admiration for the composer Johann Sebastian Bach.
  b. the role imagination plays in musical composition.
  c. that Mozart was mentally unstable.
  d. that Mozart was an imaginative child.
  e. that Mozart's only friends were imaginary people and family servants.

339. The author suggests that Mozart's adult life
  a. was ruined by repeated illness.
  b. was a disappointment after his brilliant childhood.
  c. was nothing but misery.
  d. ended in poverty and anonymity.
  e. followed the pattern of his childhood.

340. In line 57, the word *lauded* most nearly means
  a. derided.
  b. praised.
  c. punished.
  d. compensated.
  e. coveted.

341. Each of the following statements about Wolfgang Mozart is directly supported by the passage EXCEPT
  a. Mozart's father, Leopold, was instrumental in shaping his career.
  b. Wolfgang had a vivid imagination.
  c. Wolfgang's childhood was devoted to his musical career.
  d. Wolfgang's illnesses were the result of exhaustion.
  e. Maria Anna was a talented musician in her own right.

342. Based on information found in the passage, Mozart can best be described as
  a. a workaholic.
  b. a child prodigy.
  c. a sickly child.
  d. a victim of his father's ambition.
  e. the greatest composer of the eighteenth century.

Answers

299. **c.** The passage describes swing as *vibrant,* (line 6) a synonym for *lively.* It is also stated that soloists in big bands *improvised from the melody,* (line 11) indicating that the music was *melodic.*

300. **d.** In the 1940s, you would most likely hear bebop being played in clubs, such as *Minton's Playhouse in Harlem* (line 25).

301. **b.** In lines 21-22 the author states that *rhythm is the distinguishing feature of bebop.*

302. **a.** *Aficionado,* derived from the word *affection,* means a devotee or fan. The meaning can be inferred from the sentence, which states that *aficionados* flocked to clubs to *soak in the new style.* The use of *fans* in line 26 is a direct reference to the aficionados of the previous sentence.

303. **c.** The tone of the passage is neutral so only the answers beginning with *explain* or *instruct* are possible choices. The passage does not explain how to play bebop music, so **c** is the best choice.

304. **b.** Lines 2–7 describe how glam rock musicians were characterized by their flashy hair and makeup, and refers to their music as a *product,* as if it was something packaged to be sold. The choice that best describes a musician who puts outward appearance before the quality of his or her music is choice **b**, *style over substance.*

305. **c.** *Ostentatious* is an adjective that is used to describe someone or something that is conspicuously vain, or *showy.* There are numerous context clues to help you answer this question: Line 5 states that the glam rockers had a *flashy style,* and their music was *symbolic of the superficial 1980s* (line 8).

306. **d.** *Trappings* usually refer to outward decoration of dress. If you did not know the definition of trappings, the prior sentence (lines 21-24) supplies the answer: *Grunge rockers derived their fashion sense from the youth culture of the Pacific Northwest: a melding of punk rock style and outdoors clothing . . . .* The author makes no judgment of the attractiveness of grunge fashion (choice **c**).

307. **e.** Line 5 states that White Snake was a glam rock band and therefore not associated with the Seattle grunge scene. Don't be distracted by choice **a**; Mr. Epps and the Calculations may not have been a real band, but the name will nonetheless be forever associated with grunge music.

308. **b.** The relationship between grunge music and its mainstream popularity is best described as *contrary.* The most obvious example of this is found in lines 50–51, when in describing the relationship, the author states *it is very hard to buck the trend when you are the one setting it.*

309. **d.** *Ephemeral* is used to describe something that lasts only a short time, something that is *fleeting.* The context clue that best helps you to answer this question is found in lines 47–48, where the author states that grunge *faded out of the mainstream as quickly as it rocketed to prominence.*

310. **a.** If Dale was trying out equipment before it became available to the public, it makes sense that he was given original models. The passage does not specify the type of amplifiers Dale tested, so choices **c**, **d**, and **e** don't work. It specifically mentions prototypes amplifiers, so choice **b** is also wrong.

311. **c.** The clue is in the last sentence, which states that Dale was playing differently than other musicians at that time, and rock was not yet invented. Do not be distracted by the other answers, which are not supported by evidence in the passage.

312. **d.** His playing style was part of the reason the amplifiers blew up, so the answer that fits best is abrupt and disconnected. Volume was already mentioned, so you can infer that staccato does not mean loud. Line 12 holds another clue, describing his playing as a *virtual assault on the instrument.*

313. **b.** The passage is primarily about Dick Dale and his contributions to the history of playing electric guitar. The first paragraph mentions that he was called the King of Heavy Metal, and the last sentence notes that Dale made music history by playing a new kind of music that would later be called *heavy metal.*

314. **c.** To be *fazed* by something means to be disturbed or affected by it. *Unfazed* is therefore to not be affected or bothered by something. Even though the guitar was made for a right-handed player, Dale tried it anyway. He wasn't bothered by the fact that it seemingly wasn't right for him.

315. **a.** The line refers to the new transformer. Dale wanted to play louder, and that the new transformer was designed to allow him to do that. Thus, it was a *scream machine.*

316. **d.** The passage mentions that Ozzy Osbourne is often called the Father of Heavy Metal, but gives no other information about him.

317. **c.** The first paragraph *introduces* the topic of the passage, the musical traditions of colonial New Mexico. Choices **a** and **d** are too narrow, and choice **b** is too broad. Choice **e** is the purpose of the entire passage, not the first paragraph alone.

318. d. *Singular* means of or relating to a single instance, or something considered by itself. Although *strange* and *superior* can be synonyms for *singular*, the author emphasizes throughout the passage that the music is *unique*. *Unusual* is closest in meaning to unique. Also, note that in the next sentence the author states that the *bailes* had *unusual* melodic structures and the players had *unique* methods of bowing and tuning their instruments.

319. d. The passage does not explicitly state that European culture was *imposed* on native traditions. Rather, it states that the cultures mixed to give rise to the music.

320. c. The passage clearly states that both ceremonies used improvised verses. The New Year's celebration included *improvised couplets, composed on the spot* (lines 40–41) and the *verses of the song [of the wedding ceremony], played to a lively waltz, were improvised* (lines 48–49). Each of the other choices is true for one of the ceremonies but not both.

321. a. The sentence following the first mention of the ceremony states its purpose: *the community would gather to sanction the new couple* (lines 46–47). It is stated that the guests file past to *bless* the couple (line 42). Choices **b**, **d**, and **e** are all part of the ceremony but not its main purpose. Choice **c** is not explicitly supported by the text.

322. e. This title indicates that the passage covers both popular and ceremonial music and introduces the main theme of the passage: the unique (*extraordinary*) musical tradition of the Rio Grand region. The other choices are all too narrow (choice **d**), or are totally inappropriate (choice **c**).

323. b. The introductory and final paragraphs of the passage reveal the author's admiration for the music. In line 8 the author describes the musical tradition as *unique, rich* and lines 9–10 he or she calls the ceremonial music *some of the most extraordinary . . . in the Hispanic world.*" In line 62, the author describes the *indita* as *extraordinary*. Although he or she describes the tradition in positive terms, *awe* overstates the case.

324. c. In Passage 1, the author provides a limited chronology of Johnson's life (paragraphs 2, 3, and 4) and briefly describes his influence on blues and rock and roll (paragraphs 1 and 5).

325. b. In paragraph 3 of Passage 1, the author describes how Johnson was *not very good* at playing the guitar but that he *wanted to learn* and so *spent his time in blues bars watching the local blues legends* (lines 19–20). That he disappeared for some time and then returned as a *first-rate guitarist* (lines 26–27) also suggests Johnson's determination.

326. **a.** In lines 10–12 of Passage 2, the author describes how the blues came to be called the blues—thus *neologism* means a new word or new meaning or use of a word.

327. **d.** This sentence states that the blues remakes were enjoyed by all kinds of people—*black and white, young and old* (line 43)—and suggests why the songs were so popular by describing how the lyrics touched a common emotional chord in listeners, all of whom have had the blues from one or more of the sources listed in the sentence.

328. **d.** The author states that the blues was *a music perfectly suited for a nation on the brink of the Civil Rights movement* because it was music that *had the power to cross boundaries, to heal wounds, and to offer hope to a new generation of Americans* (lines 47–50). The previous sentence states that the music was popular with both the *black and white, young and old* (line 43). Thus, the author suggests that this shared musical experience helped promote understanding across racial boundaries and thereby ease racial tensions.

329. **b.** Neither author explicitly states that Robert Johnson is the best blues guitarist of his era, although this is implied by the author of Passage 1, who states that Johnson's *impact on the world of rock and roll* is *indisputable* (lines 3–4) and quotes Eric Clapton as saying Johnson is *the most important blues musician who ever lived* (lines 8–9). However, the author of Passage 2 simply lists Johnson in the same sentence as his mentors Son House and Willie Brown (lines 33–34), without suggesting that any one of these artists was better than the other.

330. **c.** Passage 1 states from the beginning that there is little information about Johnson and that the information that is available *is as much rumor as fact* (lines 2–3). There is also no definitive answer regarding how Johnson acquired his talent (paragraph 4), and the author uses the word *purportedly* in lines 34–35 to further emphasize the speculative nature of the narrative. Passage 2, on the other hand, provides many specific facts in the form of names and dates to present a text that is factual and assertive.

331. **a.** Passage 1 describes the life and influence of one specific blues artist, while Passage 2 provides a general overview of the history of the blues.

332. **c.** At the end of Passage 1, the author describes the reason so many artists record Johnson's songs: his music *capture[s] the very essence of the blues, transforming our pain and suffering with the healing magic of his guitar* (lines 41–42). This sentence "proves" the idea stated in Passage 2 that *'the blues' is something of a* *misnomer*. This is the only sentence from Passage 2 that fits the focus of Passage 1; the others concern the development or defining characteristics of the blues.

333. **a.** The passage is a neutral narration of Mozart's childhood and the beginnings of his musical career. Choices **c**, **d**, and **e** can be eliminated because the author does not take a side or try to prove a point. Choice **b** is incorrect because the author does not make any generalizations about the classical music "scene."

334. **c.** The passage clearly states that Wolfgang took an interest in the clavier when his sister was learning the instrument.

335. **c.** The passage states (lines 18–19) that Wolfgang's first *public* appearance was at Linz and that after this concert word of his genius traveled to Vienna. The passage states earlier that Vienna was the *capital* of the Hapsburg Empire.

336. **b.** The author's tone toward Leopold is *mild*—neither strongly approving nor disapproving. In a few places, however, the author conveys some disappointment, especially lines 34–36 in which she states that Leopold set an exhausting schedule for Wolfgang.

337. **c.** *Lavish* means expended or produced in abundance. Both *wasteful* and *extravagant* are synonyms for *lavish*, but, because it is modifying *palace*, *extravagant* is the more logical choice.

338. **d.** The author's language emphasizes Mozart's imagination. The phrase *engrossed in the intricacies of his make-believe court* suggests a child with a lively imagination. None of the other choices is directly supported by the text.

339. **e.** The text directly states that *the pattern established in his childhood would be the template for the rest of his short life*. Choice **d** could be misleading as the text states that Mozart was buried in an unmarked grave. However, it also states that this was customary at the time so one cannot infer that he died an anonymous pauper.

340. **b.** *Lauded* means praised or blessed. The meaning of the word can be inferred from the structure of the paragraph. The paragraph begins by summing up Mozart's childhood, and then describes how the features of his childhood were mirrored in his adult life. In his childhood Mozart *played for, and amazed, the heads of the British and French royal families* and likewise as an adult he was *lauded for his genius*. From the structure, one can infer that to be lauded is something positive. Of the positive choices, *praised* makes more sense in the sentence than *coveted*.

341. **d.** The author does not *directly* state that Mozart's illnesses were the result of exhaustion. She may imply this by describing Mozart's exhausting schedule and then stating that he became ill on tour. However, she does not make the connection explicit.

342. **b.** The main point of the passage is to describe Mozart's experiences as a child prodigy, or a highly talented child. Choices **a** and **c** are too narrow in scope, and choices **d** and **e** are not explicitly stated in the passage.

Science and Nature

Questions 343–346 are based on the following passage.

This passage is adapted from an article authored by the environmental protection organization Greenpeace, regarding Finland's destruction of old-growth forests.

(1)   Time is running out for the old-growth forests of Finland. The vast majority of Finland's valuable old-growth forest is owned by the state and logged by the state-owned company Metsähallitus. Metsähallitus' logging practices include clearcutting, logging in habitats of threat-

(5)   ened and vulnerable species, and logging in areas of special scenic or cultural value—including in areas that are critical for the reindeer herding of the indigenous Sami people.

    Despite being involved in a "dialogue process" with two environmental organizations (World Wildlife Fund and the Finnish Associa-

(10)   tion for Nature Conservation), to try and reach agreement regarding additional protection for old-growth forests, Metsähallitus is now logging sites that should be subject to negotiation.

    In June 2003, Greenpeace and the Finnish Association for Nature Conservation (FANC) presented comprehensive maps of the old-

(15)   growth areas that should be subject to moratorium, pending discussion and additional protection, to all those involved in the dialogue process. Metsähallitus then announced a halt to new logging opera-

tions in these mapped areas. Sadly, the halt in logging was short lived.
In August and September logging took place in at least six old-growth
(20)   forest areas in Northern Finland.

It seems Metsähallitus wants to have its cake and eat it too—friendly
talks with environmental groups at the same time they keep logging
critical habitat. To be blunt, their commitment to the dialog process
has proven untrustworthy. The new logging has been without con-
(25)   sensus from the dialog process or proper consultation with the Sami
reindeer herders. Now there's a risk the logging will expand to include
other old-growth areas.

Greenpeace investigations have revealed a number of companies
buying old-growth timber from Metsähallitus, but the great majority
(30)   goes to Finland's three international paper manufacturers, Stora Enso,
UPM-Kymmene, and M-Real. Greenpeace recommends that com-
panies ask for written guarantees that no material from any of the
recently mapped old-growth areas is entering or will enter their sup-
ply chain, pending the switch to only timber that has been independ-
(35)   ently certified to the standards of the Forest Stewardship Council in
order to stop this risk to protected forests.

343. According to the passage, which is NOT a logging practice
engaged in by Metsähallitus?
a. employing the clearcutting method
b. logging in the habitat of reindeer
c. logging near scenic Finnish vistas
d. logging within in the boundaries of the indigenous Sami
e. logging in traditional Norwegian Fiords

344. As used in line 15, *moratorium* most nearly means
a. an oral presentation.
b. a bipartisan meeting.
c. a cessation or stoppage.
d. an increase in volume.
e. an autopsy.

345. The author's tone may best be classified as
a. casual sarcasm.
b. urgent warning.
c. furtive anger.
d. cool indifference.
e. reckless panic.

346. The primary purpose of this passage is to
   a. alert citizens that their forests may be in danger.
   b. expose the logging industry as bad for the environment.
   c. encourage consumers to boycott Finnish wood products.
   d. agitate for change in Finland's illicit logging practices.
   e. rally support for Greenpeace international causes.

Questions 347–351 are based on the following passage.

This passage describes the Great Barrier Reef and its inhabitants.

(1)   Coral reefs are among the most diverse and productive ecosystems on
      Earth. Consisting of both living and non-living components, this type
      of ecosystem is found in the warm, clear, shallow waters of tropical
      oceans worldwide. The functionality of the reefs ranges from provid-
(5)   ing food and shelter to fish and other forms of marine life to protect-
      ing the shore from the ill effects of erosion and putrefaction. In fact,
      reefs actually create land in tropical areas by formulating islands and
      contributing mass to continental shorelines.
          Although coral looks like a plant, actually it is mainly comprised of
(10)   the limestone skeleton of a tiny animal called a coral polyp. While
      corals are the main components of reef structure, they are not the only
      living participants. Coralline algae cement the myriad corals, and
      other miniature organisms such as tube worms and mollusks con-
      tribute skeletons to this dense and diverse structure. Together, these
(15)   living creatures construct many different types of tropical reefs.
          Great Barrier Reef is the world's largest network of coral reefs,
      stretching 2,010 km (1,250 miles) off Australia's northeastern coast.
      From microorganisms to whales, diverse life forms make their home
      on the reef. Over 1,500 fish species, 4,000 mollusk species, 200 bird
(20)   species, 16 sea snake species, and six sea turtle species thrive in the
      reef's tropical waters. The reef is also a habitat for the endangered
      dugong (sea cow), moray eels, and sharks. In addition to crawling with
      animal life, the coral reef offers the viewer a spectrum of brilliant col-
      ors and intricate shapes, a virtual underwater, writhing garden.
(25)   Although protected by the Australian government, Great Barrier
      Reef faces environmental threats. Crown-of-thorns starfish feed on
      coral and can destroy large portions of reef. Pollution and rising water
      temperatures also threaten the delicate coral. But the most preventa-
      ble of the hazards to the reef are tourists. Tourists have contributed to
(30)   the destruction of the reef ecosystem by breaking off and removing
      pieces of coral to bring home as souvenirs. The government hopes

that by informing tourists of the dangers of this seemingly harmless activity they will quash this creeping menace to the fragile reef.

347. Which of the following statements does NOT describe the Great Barrier Reef?
   a. The Great Barrier reef is a colorful and active underwater structure.
   b. The Great Barrier Reef is a producer of small islands and landmasses.
   c. The Great Barrier Reef is threatened by vacationers.
   d. The Great Barrier Reef is the cause of much beachfront erosion in Northeastern Australia.
   e. The Great Barrier Reef is home to endangered sea turtles.

348. Based on information from the passage, 4,020 km would be approximately how many miles?
   a. 402
   b. 1,250
   c. 1,500
   d. 2,010
   e. 2,500

349. In line 6 of the passage, *putrefaction* most nearly means
   a. purification.
   b. decay.
   c. jettison.
   d. liquification.
   e. farming.

350. The primary purpose of this passage is to
   a. inform the reader that coral reefs are a threatened, yet broadly functioning ecosystem.
   b. alert the reader to a premier vacation destination in the tropics.
   c. explain in detail how the Great Barrier Reef is constructed.
   d. recommend that tourists stop stealing coral off the Great Barrier Reef.
   e. dispel the argument that coral is a plant, not an animal.

351. According to the passage, all of the following are a threat to a coral reef EXCEPT

a. tourists.
b. pollution.
c. erosion and putrefaction.
d. rising water temperatures.
e. Crown-of-thorns starfish.

Questions 352–358 are based on the following passage.

This passage details the history and reasoning of Daylight Saving Time.

(1)   For centuries time was measured by the position of the sun with the use of sundials. Noon was recognized when the sun was the highest in the sky, and cities would set their clock by this Apparent Solar Time, even though some cities would often be on a slightly different time. "Sum-
(5)   mer time" or Daylight Saving Time (DST) was instituted to make better use of daylight. Thus, clocks are set forward one hour in the spring to move an hour of daylight from the morning to the evening and then set back one hour in the fall to return to normal daylight.

Benjamin Franklin first conceived the idea of daylight saving during
(10)   his tenure as an American delegate in Paris in 1784 and wrote about it extensively in his essay, "An Economical Project." It is said that Franklin awoke early one morning and was surprised to see the sunlight at such an hour. Always the economist, Franklin believed the practice of moving the time could save on the use of candlelight as candles were
(15)   expensive at the time. In England, builder William Willett (1857–1915), became a strong supporter for Daylight Saving Time upon noticing blinds of many houses were closed on an early sunny morning. Willett believed everyone, including himself, would appreciate longer hours of light in the evenings. In 1909, Sir Robert Pearce
(20)   introduced a bill in the House of Commons to make it obligatory to adjust the clocks. A bill was drafted and introduced into Parliament several times but met with great opposition, mostly from farmers. Eventually, in 1925, it was decided that summer time should begin on the day following the third Saturday in April and close after the first Sat-
(25)   urday in October.

The United States Congress passed the Standard Time Act of 1918 to establish standard time and preserve and set Daylight Saving Time across the continent. This act also devised five time zones throughout the United States: Eastern, Central, Mountain, Pacific, and Alaska.
(30)   The first time zone was set on "the mean astronomical time of the sev-

enty-fifth degree of longitude west from Greenwich" (England). In 1919 this act was repealed. President Roosevelt established year-round Daylight Saving Time (also called "War Time") from 1942–1945. However, after this period each state adopted their own DST, which (35) proved to be disconcerting to television and radio broadcasting and transportation. In 1966, President Lyndon Johnson created the Department of Transportation and signed the Uniform Time Act. As a result, the Department of Transportation was given the responsibility for the time laws. During the oil embargo and energy crisis of the (40) 1970s, President Richard Nixon extended DST through the Daylight Saving Time Energy Act of 1973 to conserve energy further. This law was modified in 1986, and Daylight Saving Time was set for beginning on the first Sunday in April (to "spring ahead") and ending on the last Sunday in October (to "fall back").

(45) Through the years the U.S. Department of Transportation conducted polls concerning daylight saving time and found that many Americans were in favor of it because of the extended hours of daylight and the freedom to do more in the evening hours. In further studies the U.S. Department of Transportation also found that DST con-(50) serves energy by cutting the electricity usage in the morning and evening for lights and particular appliances. During the darkest winter months (November through February), the advantage of conserving energy in afternoon daylight saving time is outweighed by needing more light in the morning because of late sunrise. In Britain, studies (55) showed that there were fewer accidents on the road because of the increased visibility resulting from additional hours of daylight.

Despite these advantages, there is still opposition to DST. One perpetual complaint is the inconvenience of changing many clocks, and adjusting to a new sleep schedule. Farmers often wake at sunrise and (60) find that their animals do not adjust to the changing of time until weeks after the clock is either moved forward or back. In Israel, Sephardic Jews have campaigned against Daylight Saving Time because they recite prayers in the early morning during the Jewish month of Elul. Many places around the globe still do not observe day-(65) light saving time—such as Arizona (excluding Navajo reservations), the five counties in Indiana, Hawaii, Puerto Rico, Japan, and Saskatchewan, Canada. Countries located near the equator have equal hours of day and night and do not participate in Daylight Saving Time.

352. In line 20 the word *obligatory* most nearly means
   a. approved.
   b. sparse.
   c. aberrant.
   d. requisite.
   e. optional.

353. According to the passage what is the most beneficial effect of DST?
   a. changing sleeping patterns
   b. less car accidents
   c. conservation of energy
   d. additional time for family outings
   e. preferred harvesting time for farmers

354. Who first established the idea of DST?
   a. President Richard Nixon
   b. Benjamin Franklin
   c. Sir Robert Pearce
   d. President Lyndon Johnson
   e. William Willett

355. According to the passage, in which area of the world is DST least useful?
   a. the tropics
   b. Indiana
   c. Navajo reservations
   d. Mexico
   e. Saskatchewan

356. Which of the following statements is true of the U.S. Department of Transportation?
   a. It was created by President Richard Nixon.
   b. It set the standards for DST throughout the world.
   c. It constructed the Uniform Time Act.
   d. It oversees all time laws in the United States.
   e. It established the standard railway time laws.

357. What of the following statements is the best title for this passage?
- **a.** The History and Rationale of Daylight Saving Time
- **b.** Lyndon Johnson and the Uniform Time Act
- **c.** The U.S. Department of Transportation and Daylight Saving Time
- **d.** Daylight Saving Time in the United States
- **e.** Benjamin Franklin's Discovery

358. In which month does the need for more energy in the morning offset the afternoon conservation of energy by DST?
- **a.** June
- **b.** July
- **c.** October
- **d.** January
- **e.** March

Questions 359–365 are based on the following passage.

This passage details the life and illustrious career of Sir Isaac Newton,
preeminent scientist and mathematician.

(1)   Tradition has it that Newton was sitting under an apple tree when an apple fell on his head, and this made him understand that earthly and celestial gravitation are the same. A contemporary writer, William Stukeley, recorded in his *Memoirs of Sir Isaac Newton's Life* a conversa-

(5)   tion with Newton in Kensington on April 15, 1726, in which Newton recalled "when formerly, the notion of gravitation came into his mind. It was occasioned by the fall of an apple, as he sat in contemplative mood. Why should that apple always descend perpendicularly to the ground, thought he to himself. Why should it not go sideways or

(10)   upwards, but constantly to the earth's centre."
       Sir Isaac Newton, English mathematician, philosopher, and physicist, was born in 1642 in Woolsthorpe-by-Colsterworth, a hamlet in the county of Lincolnshire. His father had died three months before Newton's birth, and two years later his mother went to live with her

(15)   new husband, leaving her son in the care of his grandmother. Newton was educated at Grantham Grammar School. In 1661 he joined Trinity College, Cambridge, and continued there as Lucasian professor of mathematics from 1669 to 1701. At that time the college's teachings were based on those of Aristotle, but Newton preferred to read the

(20)   more advanced ideas of modern philosophers such as Descartes, Galileo, Copernicus, and Kepler. In 1665, he discovered the binomial

theorem and began to develop a mathematical theory that would later become calculus.

(25) However, his most important discoveries were made during the two-year period from 1664 to 1666, when the university was closed due to the Great Plague. Newton retreated to his hometown and set to work on developing calculus, as well as advanced studies on optics and gravitation. It was at this time that he discovered the Law of Universal Gravitation and discovered that white light is composed of all

(30) the colors of the spectrum. These findings enabled him to make fundamental contributions to mathematics, astronomy, and theoretical and experimental physics.

Arguably, it is for Newton's Laws of Motion that he is most revered. These are the three basic laws that govern the motion of material

(35) objects. Together, they gave rise to a general view of nature known as the clockwork universe. The laws are: (1) Every object moves in a straight line unless acted upon by a force. (2) The acceleration of an object is directly proportional to the net force exerted and inversely proportional to the object's mass. (3) For every action, there is an equal

(40) and opposite reaction.

In 1687, Newton summarized his discoveries in terrestrial and celestial mechanics in his *Philosophiae naturalis principia mathematica* (Mathematical Principles of Natural Philosophy), one of the greatest milestones in the history of science. In this work he showed how his

(45) principle of universal gravitation provided an explanation both of falling bodies on the earth and of the motions of planets, comets, and other bodies in the heavens. The first part of the *Principia*, devoted to dynamics, includes Newton's three laws of motion; the second part to fluid motion and other topics; and the third part to the system of the

(50) world, in which, among other things, he provides an explanation of Kepler's laws of planetary motion.

This is not all of Newton's groundbreaking work. In 1704, his discoveries in optics were presented in *Opticks*, in which he elaborated his theory that light is composed of corpuscles, or particles. Among his

(55) other accomplishments were his construction (1668) of a reflecting telescope and his anticipation of the calculus of variations, founded by Gottfried Leibniz and the Bernoullis. In later years, Newton considered mathematics and physics a recreation and turned much of his energy toward alchemy, theology, and history, particularly problems of

(60) chronology.

Newton achieved many honors over his years of service to the advancement of science and mathematics, as well as for his role as warden, then master, of the mint. He represented Cambridge University

in Parliament, and was president of the Royal Society from 1703 until
(65) his death in 1727. Sir Isaac Newton was knighted in 1705 by Queen
Anne. Newton never married, nor had any recorded children. He died
in London and was buried in Westminster Abbey.

359. Based on Newton's quote in lines 6–10 of the passage, what can
best be surmised about the famous apple falling from the tree?
   a. There was no apple falling from a tree—it was entirely
   made up.
   b. Newton never sits beneath apple trees.
   c. Newton got distracted from his theory on gravity by a fallen
   apple.
   d. Newton used the apple anecdote as an easily understood illustration of the Earth's gravitational pull.
   e. Newton invented a theory of geometry for the trajectory of
   apples falling perpendicularly, sideways, and up and down.

360. In what capacity was Newton employed?
   a. Physics Professor, Trinity College
   b. Trinity Professor of Optics
   c. Professor of Calculus at Trinity College
   d. Professor of Astronomy at Lucasian College
   e. Professor of Mathematics at Cambridge

361. In line 36, what does the term *clockwork universe* most nearly mean?
   a. eighteenth-century government
   b. the international dateline
   c. Newton's system of latitude
   d. Newton's system of longitude
   e. Newton's Laws of Motion

362. According to the passage, how did Newton affect Kepler's work?
   a. He discredited his theory at Cambridge, choosing to read
   Descartes instead.
   b. He provides an explanation of Kepler's laws of planetary
   motion.
   c. He convinced the Dean to teach Kepler, Descartes, Galileo, and
   Copernicus instead of Aristotle.
   d. He showed how Copernicus was a superior astronomer to
   Kepler.
   e. He did not understand Kepler's laws, so he rewrote them in
   English.

363. Which of the following is NOT an accolade received by Newton?
a. Member of the Royal Society
b. Order of Knighthood
c. Master of the Royal Mint
d. Prime Minister, Parliament
e. Lucasian Professor of Mathematics

364. Of the following, which is last in chronology?
a. *Philosophiae naturalis principia mathematica*
b. *Memoirs of Sir Isaac Newton's Life*
c. Newton's Laws of Motion
d. *Optiks*
e. invention of a reflecting telescope

365. Which statement best summarizes the life of Sir Isaac Newton?
a. distinguished inventor, mathematician, physicist, and great thinker of the seventeenth century
b. eminent mathematician, physicist, and scholar of the Renaissance
c. noteworthy physicist, astronomer, mathematician, and British Lord
d. from master of the mint to master mathematician: Lord Isaac Newton
e. Isaac Newton: founder of calculus and father of gravity

Questions 366–373 are based on the following passage.

This passage outlines the past and present use of asbestos, the potential health hazard associated with this material, and how to prevent exposure.

(1)   Few words in a contractor's vocabulary carry more negative connotations than asbestos. According to the Asbestos Network, "touted as a miracle substance," asbestos is the generic term for several naturally occurring mineral fibers mined primarily for use as fireproof insula-
(5)   tion. Known for strength, flexibility, low electrical conductivity, and resistance to heat, asbestos is comprised of silicon, oxygen, hydrogen, and assorted metals. Before the public knew asbestos could be harmful to one's health, it was found in a variety of products to strengthen them and to provide insulation and fire resistance.
(10)   Asbestos is generally made up of fiber bundles that can be broken up into long, thin fibers. We now know from various studies that when this friable substance is released into the air and inhaled into the lungs over a period of time, it can lead to a higher risk of lung cancer and a

(15)

(20)

(25)

(30)

(35)

(40)

(45)

(50)

(55)

condition known as *asbestosis.* Asbestosis, a thickening and scarring of the lung tissue, usually occurs when a person is exposed to high asbestos levels over an extensive period of time. Unfortunately, the symptoms do not usually appear until about twenty years after initial exposure, making it difficult to reverse or prevent. In addition, smoking while exposed to asbestos fibers could further increase the risk of developing lung cancer. When it comes to asbestos exposure in the home, school, and workplace, there is no safe level; any exposure is considered harmful and dangerous. Prior to the 1970s asbestos use was ubiquitous—many commercial building and home insulation products contained asbestos. In the home in particular, there are many places where asbestos hazards might be present. Building materials that may contain asbestos include fireproofing material (sprayed on beams), insulation material (on pipes and oil and coal furnaces), acoustical or soundproofing material (sprayed onto ceilings and walls), and in miscellaneous materials, such as asphalt, vinyl, and cement to make products like roofing felts, shingles, siding, wallboard, and floor tiles.

We advise homeowners and concerned consumers to examine material in their homes if they suspect it may contain asbestos. If the material is in good condition, fibers will not break down, releasing the chemical debris that may be a danger to members of the household. Asbestos is a powerful substance and should be handled by an expert. Do not touch or disturb the material—it may then become damaged and release fibers. Contact local health, environmental, or other appropriate officials to find out proper handling and disposal procedures, if warranted. If asbestos removal or repair is needed you should contact a professional.

Asbestos contained in high-traffic public buildings, such as schools presents the opportunity for disturbance and potential exposure to students and employees. To protect individuals, the Asbestos Hazard Emergency Response Act (AHERA) was signed in 1986. This law ,requires public and private non-profit primary and secondary schools to inspect their buildings for asbestos-containing building materials. The Environmental Protection Agency (EPA) has published regulations for schools to follow in order to protect against asbestos contamination and provide assistance to meet the AHERA requirements. These include performing an original inspection and periodic re-inspections every three years for asbestos containing material; developing, maintaining, and updating an asbestos management plan at the school; providing yearly notification to parent, teacher, and employee organizations regarding the availability of the school's asbestos management plan and any asbestos abatement actions taken or planned in the school; designating a contact person to ensure the responsibilities of the local education agency are properly implemented; performing periodic surveillance of known or suspected asbestos-containing building material; and providing custodial *(60)* staff with asbestos awareness training.

366. In line 12 the word *friable* most nearly means
   a. ability to freeze.
   b. warm or liquid.
   c. easily broken down.
   d. poisonous.
   e. crunchy.

367. Which title would best describe this passage?
   a. The EPA Guide to Asbestos Protection
   b. Asbestos Protection in Public Buildings and Homes
   c. Asbestos in American Schools
   d. The AHERA—Helping Consumers Fight Asbestos-Related Disease
   e. How to Prevent Lung Cancer and Asbestosis

368. According to this passage, which statement is true?
   a. Insulation material contains asbestos fibers.
   b. Asbestos in the home should always be removed.
   c. The AHERA protects private homes against asbestos.
   d. Asbestosis usually occurs in a person exposed to high levels of asbestos.
   e. Asbestosis is a man-made substance invented in the 1970s.

369. In line 23, the word *ubiquitous* most nearly means
   a. sparse.
   b. distinctive.
   c. restricted.
   d. perilous.
   e. universal.

370. Lung cancer and asbestosis are
   a. dangerous fibers.
   b. forms of serious lung disease.
   c. always fatal.
   d. only caused by asbestos inhalation.
   e. the most common illnesses in the United States.

371. The main purpose of this passage is to
   a. teach asbestos awareness in the home and schools.
   b. explain the specifics of the AHERA.
   c. highlight the dangers of asbestos to your health.
   d. provide a list of materials that may include asbestos.
   e. use scare tactics to make homeowners move to newer houses.

372. The tone of this passage is best described as
   a. cautionary.
   b. apathetic.
   c. informative.
   d. admonitory.
   e. idiosyncratic.

373. For whom is the author writing this passage?
   a. professional contractors
   b. lay persons
   c. students
   d. school principals
   e. health officials

Questions 374–381 are based on the following two passages.

The following two passages tell of geometry's Divine Proportion, 1.618.

PASSAGE 1

(1) PHI, the Divine Proportion of 1.618, was described by the astronomer
Johannes Kepler as one of the "two great treasures of geometry." (The
other is the Pythagorean theorem.)

PHI is the ratio of any two sequential numbers in the Fibonacci
(5) sequence. If you take the numbers 0 and 1, then create each subsequent number in the sequence by adding the previous two numbers,
you get the Fibonacci sequence. For example, 0, 1, 1, 2, 3, 5, 8, 13, 21,
34, 55, 89, 144. If you sum the squares of any series of Fibonacci numbers, they will equal the last Fibonacci number used in the series times
(10) the next Fibonacci number. This property results in the *Fibonacci spiral* seen in everything from seashells to galaxies, and is written math-
ematically as: $1^2 + 1^2 + 2^2 + 3^2 + 5^2 = 5 \times 8$.

Plants illustrate the Fibonacci series in the numbers of leaves, the
arrangement of leaves around the stem, and in the positioning of
(15) leaves, sections, and seeds. A sunflower seed illustrates this principal

(20) as the number of clockwise spirals is 55 and the number of counter-clockwise spirals is 89; 89 divided by 55 = 1.618, the Divine Proportion. Pinecones and pineapples illustrate similar spirals of successive Fibonacci numbers.

(20) PHI is also the ratio of five-sided symmetry. It can be proven by using a basic geometrical figure, the pentagon. This five-sided figure embodies PHI because PHI is the ratio of any diagonal to any side of the pentagon—1.618.

Say you have a regular pentagon ABCDE with equal sides and equal
(25) angles. You may draw a diagonal as line AC connecting any two vertexes of the pentagon. You can then install a total of five such lines, and they are all of equal length. Divide the length of a diagonal AC by the length of a side AB, and you will have an accurate numerical value for PHI—1.618. You can draw a second diagonal line, BC inside the pen-
(30) tagon so that this new line crosses the first diagonal at point O. What occurs is this: Each diagonal is divided into two parts, and each part is in PHI ratio (1.618) to the other, and to the whole diagonal—the PHI ratio recurs every time any diagonal is divided by another diagonal.

When you draw all five pentagon diagonals, they form a five-point
(35) star: a pentacle. Inside this star is a smaller, inverted pentagon. Each diagonal is crossed by two other diagonals, and each segment is in PHI ratio to the larger segments and to the whole. Also, the inverted inner pentagon is in PHI ratio to the initial outer pentagon. Thus, PHI is the ratio of five-sided symmetry.

(40) Inscribe the pentacle star inside a pentagon and you have the pentagram, symbol of the ancient Greek School of Mathematics founded by Pythagoras—solid evidence that the ancient Mystery Schools knew about PHI and appreciated the Divine Proportion's multitude of uses to form our physical and biological worlds.

PASSAGE 2

(1) Langdon turned to face his sea of eager students. "Who can tell me what this number is?"

A long-legged math major in back raised his hand. "That's the number PHI." He pronounced it *fee*.
(5) "Nice job, Stettner," Langdon said. "Everyone, meet PHI." [. . .] "This number PHI," Langdon continued, "one-point-six-one-eight, is a very important number in art. Who can tell me why?" [. . .] "Actually," Langdon said, [. . .] "PHI is generally considered the most beautiful number in the universe." [. . .] As Langdon loaded his slide
(10) projector, he explained that the number PHI was derived from the

Fibonacci sequence—a progression famous not only because the sum of adjacent terms equaled the next term, but because the *quotients* of adjacent terms possessed the astonishing property of approaching the number 1.618—PHI!

(15)   Despite PHI's seemingly mystical mathematical origins, Langdon explained, the truly mind-boggling aspect of PHI was its role as a fundamental building block in nature. Plants, animals, even human beings all possessed dimensional properties that adhered with eerie exactitude to the ratio of PHI to 1.

(20)   "PHI's ubiquity in nature clearly exceeds coincidence, and so the ancients assumed the number PHI must have been preordained by the creator of the universe. Early scientists heralded 1.618 as the *Divine Proportion*."

    [. . .] Langdon advanced to the next slide—a close-up of a sun-
(25) flower's seed head. "Sunflower seeds grow in opposing spirals. Can you guess the ratio of each rotation's diameter to the next?

    "1.618."

    "Bingo." Langdon began racing through slides now—spiraled pinecone petals, leaf arrangement on plant stalks, insect segmenta-
(30) tion—all displaying astonishing obedience to the Divine Proportion.

    "This is amazing!" someone cried out.

    "Yeah," someone else said, "but what does it have to do with *art*?"

    [. . .] "Nobody understood better than da Vinci the divine struc-
ture of the human body. . . . He was the first to show that the human
(35) body is literally made of building blocks whose proportional ratios
always equal PHI."

    Everyone in class gave him a dubious look.

    "Don't believe me?" . . . Try it. Measure the distance from your shoulder to your fingertips, and then divide it by the distance from
(40) your elbow to your fingertips. PHI again. Another? Hip to floor divided by knee to floor. PHI again. Finger joints. Toes. Spinal divisions. PHI, PHI, PHI. My friends, each of you is a walking tribute to the Divine Proportion." [. . . .]"In closing," Langdon said, "we return to *symbols*." He drew five intersecting lines that formed a five-pointed
(45) star. "This symbol is one of the most powerful images you will see this term. Formally known as a pentagram—or *pentacle*, as the ancients called it—the symbol is considered both divine and magical by many cultures. Can anyone tell me why that may be?"

    Stettner, the math major, raised his hand. "Because if you draw a
(50) pentagram, the lines automatically divide themselves into segments according to the Divine Proportion."

Landgon gave the kid a proud nod. "Nice job. Yes, the ratios of line segments in a pentacle *all* equal PHI, making the symbol the *ultimate* expression of the Divine Proportion."

374. The tone of Passage 2 may be described as
   a. fascinated discovery.
   b. blandly informative.
   c. passionate unfolding.
   d. droll and jaded.
   e. dry and scientific.

375. According to both passages, which of the following are synonyms?
   a. pentagon and pentacle
   b. pinecones and sunflower seed spirals
   c. Divine Proportion and PHI
   d. Fibonacci sequence and Divine Proportion
   e. Fibonacci sequence and PHI

376. In Passage 2, line 20, *ubiquity* of PHI most nearly means its
   a. rareness in nature.
   b. accuracy in nature.
   c. commonality in nature.
   d. artificiality against nature.
   e. purity in an unnatural state.

377. Both passages refer to the "mystical mathematical" side of PHI. Based on the two passages, which statement is NOT another aspect of PHI?
   a. PHI is a ratio found in nature.
   b. PHI is the area of a regular pentagon.
   c. PHI is one of nature's building blocks.
   d. PHI is derived from the Fibonacci sequence.
   e. PHI is a math formula.

378. Which of the following techniques is used in Passage 1, lines 13–18 and Passage 2, lines 24–26?
   a. explanation of terms
   b. comparison of different arguments
   c. contrast of opposing views
   d. generalized statement
   e. illustration by example

379. All of the following questions can be explicitly answered on the basis of the passage EXCEPT
   a. What is the ratio of the length of one's hip to floor divided by knee to floor?
   b. What is the precise mathematical ratio of PHI?
   c. What is the ratio of the length of one's shoulder to fingertips divided by elbow to fingertips?
   d. What is the ratio of the length of one's head to the floor divided by shoulder's to the floor?
   e. What is the ratio of each sunflower seed spiral rotation's diameter to the next?

380. According to both passages, the terms *ancient Mystery Schools* (Passage 1, line 43), *early scientists* (Passage 2, line 22), and *ancients* (Passage 2, line 46) signify what about the divine proportion?
   a. Early scholars felt that the Divine Proportion was a magical number.
   b. Early scholars found no scientific basis for the Divine Proportion.
   c. Early mystery writers used the Divine Proportion.
   d. Early followers of Pythagoras favored the Pythagorean theorem over the divine proportion.
   e. Early followers of Kepler used the Divine Proportion in astronomy.

381. Which of the following is NOT true of the pentagon?
   a. It is considered both divine and magical by many cultures.
   b. It is a geometric figure with five equal sides meeting at five equal angles.
   c. It is a geometric figure whereby PHI is the ratio of any diagonal to any side.
   d. If you draw an inverted inner pentagon inside a pentagon, it is in PHI ratio to the initial outer pentagon.
   e. A polygon having five sides and five interior angles is called a pentagon.

Questions 382–390 are based on the following passage.

The following passage describes the composition and nature of ivory.

(1)   Ivory skin, ivory teeth, Ivory Soap, Ivory Snow—we hear "ivory" used all the time to describe something fair, white, and pure. But where does ivory come from, and what exactly is it? Is it natural or man-

(5) made? Is it a modifier, meaning something pure and white or is it a specialized and discrete substance?

Historically, the word *ivory* has been applied to the tusks of ele-phants. However, the chemical structure of the teeth and tusks of mammals is the same regardless of the species of origin, and the trade in certain teeth and tusks other than elephant is well established and (10) widespread. Therefore, ivory can correctly be used to describe any mammalian tooth or tusk of commercial interest that is large enough to be carved or scrimshawed. Teeth and tusks have the same origins. Teeth are specialized structures adapted for food mastication. Tusks, which are extremely large teeth projecting beyond the lips, have (15) evolved from teeth and give certain species an evolutionary advantage that goes beyond chewing and breaking down food in digestible pieces. Furthermore, the tusk can be used to actually secure food through hunting, killing, and then breaking up large chunks of food into manageable bits.

(20) The teeth of most mammals consist of a root as well as the tusk proper. Teeth and tusks have the same physical structures: pulp cavity, dentine, cementum, and enamel. The innermost area is the pulp cav-ity. The pulp cavity is an empty space within the tooth that conforms to the shape of the pulp. Odontoblastic cells line the pulp cavity and (25) are responsible for the production of dentine. Dentine, which is the main component of carved ivory objects, forms a layer of consistent thickness around the pulp cavity and comprises the bulk of the tooth and tusk. Dentine is a mineralized connective tissue with an organic matrix of collagenous proteins. The inorganic component of dentine (30) consists of dahllite. Dentine contains a microscopic structure called dentinal tubules which are micro-canals that radiate outward through the dentine from the pulp cavity to the exterior cementum border. These canals have different configurations in different ivories and their diameter ranges between 0.8 and 2.2 microns. Their length is (35) dictated by the radius of the tusk. The three dimensional configura-tion of the dentinal tubules is under genetic control and is therefore a characteristic unique to the order of the mammal.

Exterior to the dentine lies the cementum layer. Cementum forms a layer surrounding the dentine of tooth and tusk roots. Its main func-(40) tion is to adhere the tooth and tusk root to the mandibular and max-illary jaw bones. Incremental lines are commonly seen in cementum.

Enamel, the hardest animal tissue, covers the surface of the tooth or tusk which receives the most wear, such as the tip or crown. Ameloblasts are responsible for the formation of enamel and are lost (45) after the enamel process is complete. Enamel exhibits a prismatic structure with prisms that run perpendicular to the crown or tip. Enamel prism patterns can have both taxonomic and evolutionary significance.

Tooth and tusk ivory can be carved into an almost infinite variety of shapes and objects. A small example of carved ivory objects are small (50) statuary, netsukes, jewelry, flatware handles, furniture inlays, and piano keys. Additionally, wart hog tusks, and teeth from sperm whales, killer whales, and hippos can also be scrimshawed or superficially carved, thus retaining their original shapes as morphologically recognizable objects.

The identification of ivory and ivory substitutes is based on the (55) physical and chemical class characteristics of these materials. A common approach to identification is to use the macroscopic and microscopic physical characteristics of ivory in combination with a simple chemical test using ultraviolet light.

382. In line 5, what does the term *discrete* most nearly mean?
  a. tactful
  b. distinct
  c. careful
  d. prudent
  e. judicious

383. Which of the following titles is most appropriate for this passage?
  a. Ivory: An Endangered Species
  b. Elephants, Ivory, and Widespread Hunting in Africa
  c. Ivory: Is It Organic or Inorganic?
  d. Uncovering the Aspects of Natural Ivory
  e. Scrimshaw: A Study of the Art of Ivory Carving

384. The word *scrimshawed* in line 12 and line 52 most nearly means
  a. floated.
  b. waxed.
  c. carved.
  d. sunk.
  e. buoyed.

385. Which of the following choices is NOT part of the physical structure of teeth?
  a. pulp cavity
  b. dentine
  c. cementum
  d. tusk
  e. enamel

386. As used in line 13, what is the best synonym for *mastication*?
a. digestion
b. tasting
c. biting
d. chewing
e. preparation

387. Which sentence best describes *dentinal tubules*?
a. Dentinal tubules are a layer surrounding the dentine of tooth and tusk roots.
b. Dentinal tubules are micro-canals that radiate outward through the dentine from the pulp cavity to the exterior cementum border.
c. Dentinal tubules are responsible for the formation of enamel and are lost after the enamel process is complete.
d. Dentinal tubules cover the surface of the tooth or tusk which receives the most wear, such as the tip or crown.
e. Dentinal tubules are extremely large teeth projecting beyond the lips that have evolved from teeth and give certain species an evolutionary advantage.

388. According to the passage, all of the following are organic substances EXCEPT
a. cementum.
b. dentine.
c. dahllite.
d. ameloblasts.
e. collagen.

389. According to the passage, how can natural ivory be authenticated?
a. by ultraviolet light
b. by gamma rays
c. by physical observation
d. by osmosis
e. by scrimshaw

390. According to the passage, which statement is NOT true of enamel?
a. It is an organic substance.
b. It is the hardest of animal tissues.
c. It should never be exposed to ultraviolet light.
d. It structure is prismatic.
e. It is formed with the aid of ameloblasts.

Questions 391–399 are based on the following passage.

This passage is about the process by which scientists prove theories, the scientific method.

(1)   The scientific method usually refers to either a series or a collection of processes that are considered characteristic of scientific investigation and of the acquisition of new scientific knowledge.
The essential elements of the scientific method are:

(5)   *Observe:* Observe or read about a phenomenon.
Hypothesize: Wonder about your observations, and invent a hypothesis, or a guess, which could explain the phenomenon or set of facts that you have observed.
Test: Conduct tests to try out your hypothesis.
(10)   *Predict:* Use the logical consequences of your hypothesis to predict observations of new phenomena or results of new measurements.
Experiment: Perform experiments to test the accuracy of these predictions.
(15)   *Conclude:* Accept or refute your hypothesis.
Evaluate: Search for other possible explanations of the result until you can show that your guess was indeed the explanation, with confidence.
Formulate new hypothesis: as required.

(20)   This idealized process is often misinterpreted as applying to scientists individually rather than to the scientific enterprise as a whole. Science is a social activity, and one scientist's theory or proposal cannot become accepted unless it has been published, peer reviewed, criticized, and finally accepted by the scientific community.

(25)   **Observation**
The scientific method begins with observation. Observation often demands careful *measurement*. It also requires the establishment of an *operational definition* of measurements and other concepts before the experiment begins.

(30)   **Hypothesis**
To explain the observation, scientists use whatever they can (their own creativity, ideas from other fields, or even systematic guessing) to come up with possible explanations for the phenomenon under

(35) study. Deductive reasoning is the way in which predictions are used to test a hypothesis.

Testing

In the twentieth century, philosopher Karl Popper introduced the idea that a hypothesis must be falsifiable; that is, it must be capable of being demonstrated wrong. A hypothesis must make specific predictions;
(40) these predictions can be tested with concrete measurements to support or refute the hypothesis. For instance, Albert Einstein's theory of general relativity makes a few specific predictions about the structure of space and flow of time, such as the prediction that light bends in a strong gravitational field, and the amount of bending depends in a pre-
(45) cise way on the strength of the gravitational field. Observations made of a 1919 solar eclipse supported this hypothesis against other possible hypotheses, such as Sir Isaac Newton's theory of gravity, which did not make such a prediction. British astronomers used the eclipse to prove Einstein's theory and therefore, eventually replaced Newton's
(50) theory.

Verification

Probably the most important aspect of scientific reasoning is verification. Verification is the process of determining whether the hypothesis is in accord with empirical evidence, and whether it will
(55) continue to be in accord with a more generally expanded body of evidence. Ideally, the experiments performed should be fully described so that anyone can reproduce them, and many scientists should independently verify every hypothesis. Results that can be obtained from experiments performed by many are termed *reproducible* and are
(60) given much greater weight in evaluating hypotheses than non-reproducible results.

Evaluation

Falsificationism argues that any hypothesis, no matter how respected or time-honored, *must* be discarded once it is contradicted by new reli-
(65) able evidence. This is, of course, an oversimplification, since individual scientists inevitably hold on to their pet theory long after contrary evidence has been found. This is not always a bad thing. Any theory can be made to correspond to the facts, simply by making a few adjustments—called "auxiliary hypothesis"—so as to bring it into corre-
(70) spondence with the accepted observations. The choice of when to reject one theory and accept another is inevitably up to the individual scientist, rather than some methodical law.

Hence *all* scientific knowledge is always in a state of flux, for at any time new evidence could be present[ed] that contradicts long-held
(75) hypotheses.

The experiments that reject a hypothesis should be performed by many different scientists to guard against bias, mistake, misunderstanding, and fraud. Scientific journals use a process of *peer review*, in which scientists submit their results to a panel of fellow scientists (who may or
(80) may not know the identity of the writer) for evaluation. Peer review may well have turned up problems and led to a closer examination of experimental evidence for many scientists. Much embarrassment, and wasted effort worldwide, has been avoided by objective peer review, in addition to continuing the use and proving the necessity of the scientific method.

391. Which step in the process of scientific method do lines 63–72 speak of?
   a. operational definition
   b. verification
   c. evaluation
   d. phenomenon
   e. hypothesizing

392. What is the tone of this passage?
   a. enigmatic
   b. apathetic
   c. abstruse
   d. instructive
   e. revealing

393. In line 63 the word *falsificationism* most nearly means
   a. validation.
   b. qualification.
   c. confirmation.
   d. facilitation.
   e. refutation.

394. Which statement is FALSE?
   a. Reproducible results can be obtained by experiments performed by a variety of scientists.
   b. An auxiliary hypothesis can be made to correspond to the facts.
   c. Einstein's theory of relativity makes space and time predictions.
   d. Peer review is usually not a valuable tool for scientists.
   e. Experiments are a necessary element in the scientific method.

395. According to the passage, which is true of a hypothesis?
  a. It is not a necessary process in the scientific method.
  b. It cannot be discarded by a competing theory.
  c. It is a guess.
  d. It can make a broad and general prediction.
  e. It is always considered auxiliary.

396. What is the best title for this passage?
  a. The Theory of Relativity
  b. The Scientific Method: A Step-by-Step Process
  c. The Two Stages of Proving Theories
  d. How to Form a Hypotheses
  e. Evaluating Data with the Scientific Method

397. What is meant by the term *operational definition* in line 28
  of the passage?
  a. a scientific law
  b. a theory
  c. a clear definition [of a measurement]
  d. scientific method
  e. hypothesis

398. What do lines 37–48 of the passage indicate?
  a. The theory of general relativity is a hypothesis.
  b. Karl Popper proved the theory of relativity to be incorrect.
  c. Einstein was the father of the scientific method.
  d. Space and the flow of time theories are still in a state of flux.
  e. Sir Isaac Newton's theory of gravity disproved Einstein's theory.

399. Which is NOT a step used in the process of scientific method?
  a. observation
  b. simplification
  c. evaluation
  d. verification
  e. hypothesize

Answers

343. **e.** Answer choices **a–d** are all unauthorized logging practices performed by Metsähallitus in Finland. Choice **e** is incorrect because it refers to another country.

344. **c.** Calling for a moratorium means to cease or stop an activity or concept. You can deduce this correct answer from the clue in line 17, *halt*.

345. **b.** The author's tone can best be characterized as an *urgent warning*. The passage exposes an illegal logging practice that threatens to destroy forests in Finland. The author's genuine concern rules out choices **a** and **d**, and there is nothing in the passage to suggest that the author is either secretly angry, choice **c**, or in a state of panic, choice **e**.

346. **d.** Though Greenpeace is clearly out to inform the reader of bad logging practices in Finland, it is not trying to rally support for their organization (choice **e**); rather, their goal is to promote awareness, and through awareness, change. Choice **a**'s suggestion that other forests are endangered is false, and choices **b** and **c** are not ideas put forth by the passage.

347. **d.** The Great Barrier Reef does not cause erosion; it prevents it. All of the other choices are true and can be found in the passage.

348. **e.** According to the passage, 2,010 km is approximately 1,250 miles. So, twice as many km (4,020) would be approximately twice as many miles (2,500).

349. **b.** The phrase *ill effects of* that precedes the words *erosion* and *putrefaction* means that putrefaction is a negative consequence, as is erosion. The other choices are either neutral, **c**, **d**, and **e**, or positive, **a**.

350. **a.** This statement encapsulates the entire passage, not just a part of it. Choices **c** and **e** are too specific to be correct. Choices **b** and **d** are not supported by the passage.

351. **c.** *Erosion and putrefaction* (line 6) are the consequences to shoreline if the coral reefs are neglected or destroyed.

352. **d.** Choices **b** and **c**, meaning scattered and erratic respectively, are not supported in the document. Choice **e** is incorrect because it is an antonym of *obligatory*. Choice **a** may be considered a synonym but it is not the best choice. The best choice is **d**, *requisite*.

353. **c.** This answer can be drawn from lines 48–51 in the passage. Choices **b** and **d** are also true but not the best answers. Choices **a** and **e** are not true.

354. **b.** Lines 9–11 of the passage clearly state that Benjamin Franklin first considered the concept of DST.

355. **a.** Locations near the equator do not participate in DST because they have equal hours of day and night; therefore, DST, which extends the daylight period, is not useful. Choice **c** is incorrect because Navajo reservations observe DST. Choice **b** is incorrect because parts of Indiana do observe DST. Choice **d** is incorrect because Mexico now observes DST. Choice **e** is incorrect because Saskatchewan chooses to not observe DST.

356. **d.** This choice is directly supported by lines 37–39 of the passage.

357. **a.** Choices **b**–**e** are incorrect because they each refer to specific points raised in the passage, but not throughout the passage. Only choice **a** describes the point of the entire passage.

358. **d.** This choice is directly supported by lines 51–54 of the passage.

359. **d.** The anecdote contrasts with the ensuing quote in paragraph 1 and depicts a plausible reason for the apple story—Newton wanted to make his theory understood to the general public. Speaking in physics terminology is abstract, but using an illustration that regular people have witnessed again and again would aid in understanding. The quote gives credence to the anecdote, ruling out choice **a**. Choices **b** and **e** are never mentioned, and choice **c** is not backed up by the passage.

360. **e.** Lines 16–18 of the passage clearly state that Newton became Professor of Mathematics at Trinity College, Cambridge.

361. **e.** In paragraph 4, Newton's Laws of Motion are said to *govern the motion of objects* and are the basis for the concept of the *clockwork universe*. Nowhere in the passage is it stated that Newton or his Laws are responsible for the international dateline (choice **b**), latitude (choice **c**), or longitude (choice **d**). Choice **a** plays on the word *govern* in line 34 and is misleading.

362. **b.** Lines 49–51 specifically state that Newton provided an explanation of Kepler's laws.

363. **d.** All of the other titles were bestowed on Newton during his lifetime.

364. **b.** William Stukeley published *Memoirs of Sir Isaac Newton's Life* in 1726, after Newton's death. The other choices are all accomplishemnts of Newton in his lifetime.

365. **a.** Choice **a** is correct because it lists the proper accolades and the proper timeframe in which he lived. Choice **b** is incorrect because he did not live in the Renaissance; choices **c** and **d** are incorrect because he was not a lord, but a knight; and choice **e** is incorrect because it is not the best summary of his vast accomplishments.

366. **c.** The phrase *broken up into long, thin fibers* is used to describe asbestos bundles in lines 10–11, prior to the word phrase *friable substance* supports that friable means easily broken down. All other choices are not supported in the passage.

367. **b.** This choice best describes the passage in its entirety, while the other choices describe individual points made throughout the passage.

368. **d.** Asbestosis usually occurs in people exposed to high levels of asbestos. Choice **a** is incorrect because not all insulation material contains asbestos fibers; choice **b** is incorrect because asbestos that is in good condition and not crumbled or breaking away does not need to be removed. Choice **c** is incorrect because the AHERA protects schools against asbestos exposure. Choice **e** is incorrect because asbestosis is a lung disease not a manmade substance.

369. e. The correct choice is universal. The sentence *Many commercial building and home insulation products contained asbestos* after the word *ubiquitous* shows that asbestos was commonly used.

370. **b.** The author explains that lung cancer and asbestosis are diseases of the lung in lines 11–16 of the passage. Choice **a** is not true because lung cancer and asbestosis are not dangerous fibers, asbestos is. Choice **c** is incorrect as both diseases may be fatal, but may be treated, as well. Choice **d** is incorrect because we know lung cancer can develop in ways other than asbestos exposure. Choice **e** is incorrect because asbestosis in not necessarily a common illness.

371. **a.** While the passage does include the other choices except choice **e**, the overall purpose of the passage is to teach asbestos awareness in the home and school.

372. **c.** The tone of this passage is informative, serving to instruct the general reader about asbestos. Choices **a** and **d** (*cautionary* and *admonitory*) are synonyms, and while the passage does show the dangers of asbestos, the general tone is not cautionary. *Apathetic* (choice **b**) means indifferent and *idiosyncratic* (choice **e**) means distinctive, neither of which apply.

373. **b.** The author is writing for a layperson, meaning a homeowner, parent, or student. Choices **a** and **e** describe professionals, while **c** and **d** describe people you would find in a school setting, all of whom may be interested in this information, but none of whom is the specific, targeted audience.

374. **a.** The tone is best described as one of fascinated discovery unfolding during a lecture. A clue to the tone is the use of exclamation points and the excited, choppy delivery of Langdon's information. Choices **b** or **e** may be considered as the tone of Passage 1. Choice **c** and **d** are not supported by the text.

375. c. The mathematical ratio PHI is also known as the Divine Proportion. This is directly stated in lines 17–18 of Passage 1, and lines 22–23 of Passage 2.

376. c. *Ubiquity* is used here to show that even though the concept of PHI in nature seems unusual or unique at first, it is actually a very common and predictable occurrence. The other choices are not supported by the passage.

377. b. PHI is not the area of a regular pentagon. All other answers describe an aspect of PHI as found in the two passages.

378. e. The subject of both paragraphs is Fibonacci spirals. Sunflower seeds, pinecones, and pineapples are mentioned as examples of the Fibonacci spiral.

379. d. The answer for choices **a**, **b**, **c**, and **e** are all the same, according to Passage 2: 1.618. The ratio of head to floor divided by shoulder to floor (choice **d**) is not covered in the passage.

380. a. Both passage refer to the fact that early or ancient scientists perceived the Divine Proportion to be a magical number. Choices **d** and **e** could be correct, but they are not supported by the passage. Choices **b** and **c** are false.

381. a. This statement, while true, refers to the pentagram, not the pentagon. Choices **b–e** are all true statements about the pentagon.

382. b. *Discrete* means *distinct*, and as used in the passage, it is paired with *specialized*, a context clue. Choices **a**, **c**, **d**, and **e** are all synonyms for the homophone, *discreet*.

383. d. Choice **b** is not covered in the passage. Choices **a**, **c**, and **e**, while mentioned, are too specific to be viable titles. Choice **d** is broad-ranging enough to encompass the entire passage.

384. c. *Scrimshawed* means carved, as in line 12. The word is often associated with whaling and seafaring, so answer choices **a**, **d**, and **e** are all distracters stemming from that confusion regarding context. Because scrimshaw and enamel are wax-like substances, a less careful reader may choose **b**.

385. d. According to lines 21–22 of the passage, choices **a**, **b**, **c**, and **e** are all parts of the physical structure of teeth. Choice **d**, *tusk*, is not a component of teeth, but rather a type of tooth found in some mammals.

386. d. From the context in lines 13–17, it can be deduced that *mastication* means the act of chewing because tusks, evolved from teeth,

are described in line 16 as able to go *beyond chewing.* Choices **a**,
b, and **c** are distracters that might be chosen if not reading carefully. Choice **e**, *preparation*, is too vague.

387. b. Lines 30–32 clearly state that dentinal tubules *are micro-canals
that radiate outward through the dentine from the pulp cavity to the
exterior cementum border.*

388. c. In the passage, the substances in choices **a**, **b**, **d**, and **e** are all
described as organic substances. Therefore, choice **c**, an inor-
ganic substance, is correct.

389. a. Lines 55–58 identify how natural ivory can be authenticated.

390. c. According to the fifth paragraph of the passage, enamel is the
hardest animal tissue (animal tissue, by nature, is a living thing,
and thus organic), ameloblasts help form it, and it has a prismatic structure (choices **a**, **b**, **d**, and **e**). Choice **c** is incorrect
because lines 55–58 state that ivory is commonly tested via
ultraviolet light, which would indicate exposure.

391. c. Choice **c** is correct because these lines specifically speak to the
evaluation process of the scientific method.

392. d. The entire passage is instructive and about educating
the reader.

393. e. *Falsificationism* means to refute and prove wrong as supported in
lines 38 and 63 of the passage.

394. d. *Peer review* is proposed as a vital part of the scientific method,
and it is directly supported as such by lines 61–67 in the passage. The other statements are all true.

395. c. Lines 31–35 of the passage support this truth about hypotheses.
The other statements about hypotheses are false.

396. b. This is the best choice as it explains the overall point of the passage, which is a step-by-step process covering the scientific
method. Choice **e** is close, but the entire passage is not about
evaluating data. Choice **a** is incorrect because the theory of relativity is only cited as an example, not as a general topic. Likewise, choice **c** only considers a small part of the passage. Choice
d is too specific.

397. c. *Operational definition* is defined as a clear definition of a measurement in lines 26–29 in the passage.

398. a. Choice **a** is supported by the passage. Choice **c** is not supported
anywhere in the passage. Choices **b**, **d**, and **e** are all incorrect
interpretations of information contained in the passage and are
careless choices.

399. b. All the other choices are indicated in the passage to be steps of
the process of scientific method.

Sports and Leisure

Questions 400–402 are based on the following passage.

In the following passage, the author attempts to define what separates a sport from a leisure activity.

(1)   The seemingly simple question of *"what defines a sport?"* has been the fodder for argument and conversation for years, among professional and armchair athletes alike. There seems to be no doubt that vigorous and highly competitive activities such as baseball, football, and

(5)   soccer are truly "sports," but when the subject of other activities such as darts, chess, and shuffleboard is broached we find ourselves at the heart of a controversy.

If say, billiards, is not a sport, then what exactly is it? Those who would dispute it to be a sport would respond that it is a simple leisure

(10)   activity. They would go on to claim a true sport first and foremost requires some form of physical exertion. More to the point, if a player does not break a sweat, what he or she plays is not a sport. Beyond that, more important criteria would be the need for decent hand-eye coordination, and the ever-present possibility of sustaining injury. Bil-

(15)   liards only fits one of those specifications (hand-eye coordination), so according to the doubters, it is not a real sport.

To help resolve this dispute, the first text to consult would have to be the dictionary. According to one dictionary, a sport is defined as "a

(20) diversion" or a "recreation." Assuming one strictly adheres to the simple guidelines laid out in that definition, it would seem that almost any activity that provides enjoyment could be classified as a sport. And if, according to the dictionary, watching a sport on television is a sport itself, I guess that would make a couch potato an athlete. Play ball!

400. The author's tone in this passage could be described as
a. serious.
b. light-hearted.
c. confrontational.
d. dark.
e. romantic.

401. The word *vigorous* in line 3 most nearly means
a. languorous.
b. boring.
c. intricate.
d. ancient.
e. strenuous.

402. According to the criteria given in lines 11–14, all of the following would be considered a "true" sport EXCEPT
a. cheerleading.
b. skiing.
c. race car driving.
d. horse shoes.
e. gymnastics.

Questions 403–407 are based on the following passage.

The following passage describes the Native American games that were predecessors to the modern sport of lacrosse.

(1) The roots of the modern-day sport of lacrosse are found in tribal stick and ball games developed and played by many native North American tribes dating back as early as the fifteenth century. The Native American names for these games reflected the bellicose nature of those early (5) contests, many of which went far beyond friendly recreational competition. For example, the Algonquin called their game *Baggattaway*, which meant, "they bump hips." The Cherokee Nation and the Six Tribes of the Iroquois called their sport *Tewaarathon*, which translated into "Little Brother of War." Rules and style of play differed from

(10)   tribe to tribe and games could be played by as few as fifteen to as many as 1,000 men and women at a time. These matches could last for three days, beginning at dawn each day and ending at sunset. The goals could be specific trees or rocks, and were a few hundred yards to a few miles apart. Despite these differences, the sole object of every game

(15)   was the same: to score goals by any means necessary. Serious injuries caused by blows from the heavy wooden sticks used in the games were not uncommon, and often expected. Not surprisingly, the Native Americans considered these precursors to today's lacrosse excellent battle preparation for young warriors, and games were often used to

(20)   settle disputes between tribes without resorting to full-blown warfare.

For the Six Tribes of the Iroquois, certain matches of *Tewaarathon* held religious significance, as well. One of the most important gods the Iroquois worshipped was the Creator, *Deganawidah*. In Iroquois legend, the Creator united the Six Tribes into the one nation.

(25)   *Tewaarathon* was played to please the Creator, and the competition was viewed as a recreation of the Iroquois Creation Story, where supernatural forces of good and evil battled each other in an epic struggle.

403. In line 4, *bellicose* most closely means
   a. beautiful.
   b. warlike.
   c. peaceful.
   d. family minded.
   e. clumsy.

404. The passage describes the early versions of lacrosse as
   a. strictly regulated competitions.
   b. intense games played against the Pilgrims.
   c. serious and meaningful matches.
   d. played only by the best athletes selected from each tribe.
   e. friendly exhibitions.

405. Which of the following titles would be the most appropriate for this passage?
   a. Little Brother of War
   b. Lacrosse: America's Most Violent Sport
   c. The Origins of the Modern Lacrosse Stick
   d. Deganawidah and the Six Tribes
   e. Hockey: the Little Brother of Lacrosse

406. In line 15, the author's use of the phrase *by any means necessary* emphasizes the

    a. unpredictable nature of the game.

    b. mild nature of the game.

    c. violent nature of the game.

    d. fact that both women and men participated in the games.

    e. importance of scoring goals.

407. The author's main purpose for writing this passage is to

    a. illustrate the differences between the early games and today's lacrosse.

    b. condemn the violent tactics often used by the Native American players.

    c. show how ancient games influenced many games played today.

    d. teach the reader about the Iroquois Creation Story.

    e. describe the importance of these games in Native American culture.

Questions 408–412 are based on the following passage.

The following passage is adapted from a critical commentary about commercialism in today's society.

(1)   Traditional body signage seems largely to have disappeared. Well, many of the old symbols and names are still around, of course, but they are part of the commercial range of options. Seeing someone in a Harvard or Oxford sweatshirt or a kilt or a military tie now com-

(5)   municates nothing at all significant about that person's life other than the personal choice of a particular consumer. Religious signs are still evocative, to be sure, but are far less common than they used to be. Why should this be? I suspect one reason may be that we have lost a sense of significant connection to the various things indicated by such

(10)   signs. Proclaiming our high school or university or our athletic team or our community has a much lower priority nowadays, in part because we live such rapidly changing lives in a society marked by constant motion that the stability essential to confer significance on such signs has largely gone.

(15)   But we still must attach ourselves to something. Lacking the conviction that the traditional things matter, we turn to the last resort of the modern world: the market. Here there is a vast array of options, all equally meaningless in terms of traditional values, all equally important in identifying the one thing left to us for declaring our

(20) identity publicly, our fashion sense and disposable income. The market naturally manipulates the labels, making sure we keep purchasing what will most quickly declare us excellent consumers. If this year a Chicago Bulls jacket or Air Jordan shoes are so popular that we are prepared to spend our way into a trendy identity, then next year there
(25) will be something else.

408. The main purpose of the passage is to
   a. discuss basketball's importance in today's fashions.
   b. relate the tribal history of tattoos.
   c. tell a story about the good old days.
   d. help the reader discover his or her own true identity.
   e. discuss commercialism's powerful influence upon personal identity.

409. What does the author mean by the *commercial range of options* (line 3)?
   a. the variety of commercials on television and radio
   b. the numerous products available to today's consumer
   c. the ability to shop on the Internet
   d. let the buyer beware
   e. technology's impact upon the world

410. In line 20, *disposable income* refers to
   a. recyclable goods.
   b. spending money.
   c. life savings.
   d. a donation to charity.
   e. garbage.

411. The author would agree with all the following statements EXCEPT
   a. A person wearing a New York Yankees baseball hat is not necessarily a fan of the team or a resident of New York.
   b. Pride in our school or community is not as strong today as it was years ago.
   c. In today's society, being trendy is more important than keeping tradition.
   d. You can tell a lot about somebody by what they are wearing.
   e. The last resort of the modern world is the marketplace.

412. Which statement best simplifies the author's point of view of today's society in lines 12–14?
 a. Times have changed.
 b. People's lives today are very similar to those of a generation ago.
 c. Fashion is very important in today's world.
 d. People today don't have proper nutrition.
 e. Life is short.

Questions 413–421 are based on the following passage.

The following passage is an excerpt from Jack London's **The Cruise of the Snark**. *In this selection, London discusses his experience of learning to surf in Waikiki in the early 1900s.*

(1) A wave is a communicated agitation. The water that composes the body of a wave does not move. If it did, when a stone is thrown into a pond and the ripples spread away in an ever-widening circle, there would appear at the center an ever-increasing hole. No, the water that *(5)* composes the body of a wave is stationary. Thus, you may watch a particular portion of the ocean's surface and you will see the same water rise and fall a thousand times to the agitation communicated by a thousand successive waves. Now imagine this communicated agitation moving shoreward. As the bottom shoals, the lower portion of the *(10)* wave strikes land first and is stopped. But water is fluid, and the upper portion has not struck anything, wherefore it keeps on communicating its agitation, keeps on going. And when the top of the wave keeps on going, while the bottom of it lags behind, something is bound to happen. The bottom of the wave drops out from under and the top of *(15)* the wave falls over, forward, and down, curling and cresting and roaring as it does so. It is the bottom of a wave striking against the top of the land that is the cause of all surfs.

But the transformation from a smooth undulation to a breaker is not abrupt except where the bottom shoals abruptly. Say the bottom *(20)* shoals gradually from a quarter of a mile to a mile, then an equal distance will be occupied by the transformation. Such a bottom is that off the beach of Waikiki, and it produces a splendid, surf-riding surf. One leaps upon the back of a breaker just as it begins to break, and stays on it as it continues to break all the way in to shore.

(25) And now to the particular physics of surf-riding. Get out on a flat board, six feet long, two feet wide, and roughly oval in shape. Lie down upon it like a small boy on a coaster and paddle with your hands

out to deep water, where the waves begin to crest. Lie out there quietly on the board. Sea after sea breaks before, behind, and under and
(30) over you, and rushes in to shore, leaving you behind. When a wave
crests, it gets steeper. Imagine yourself, on your board, on the face of
that steep slope. If it stood still, you would slide down just as a boy
slides down a hill on his coaster. "But," you object, "the wave doesn't
stand still." Very true, but the water composing the wave stands still,
(35) and there you have the secret. If ever you start sliding down the face
of that wave, you'll keep on sliding and you'll never reach the bottom.
Please don't laugh. The face of that wave may be only six feet, yet you
can slide down it a quarter of a mile, or half a mile, and not reach the
bottom. For, see, since a wave is only a communicated agitation or
(40) impetus, and since the water that composes a wave is changing every
instant, new water is rising into the wave as fast as the wave travels.
You slide down this new water, and yet remain in your old position
on the wave, sliding down the still newer water that is rising and
forming the wave. You slide precisely as fast as the wave travels. If it
(45) travels fifteen miles an hour, you slide fifteen miles an hour. Between
you and shore stretches a quarter of mile of water. As the wave travels, this water obligingly heaps itself into the wave, gravity does the
rest, and down you go, sliding the whole length of it. If you still cher-
ish the notion, while sliding, that the water is moving with you, thrust
(50) your arms into it and attempt to paddle; you will find that you have to
be remarkably quick to get a stroke, for that water is dropping astern
just as fast as you are rushing ahead.

413. The author compares surfing to
    a. an ever-increasing hole forming in the water.
    b. a chemistry experiment gone wrong.
    c. a boy sledding down a hill on a coaster.
    d. a transformation of time and space.
    e. flying through the air like a bird.

414. All of the following questions can be answered based on
information from the passage EXCEPT
    a. When a wave crests, it gets steeper.
    b. If a wave is moving at eight miles per hour, so is the surfer on
that wave.
    c. A wave is constantly recomposing itself with new water.
    d. A flat board is the most popular type of surfboard.
    e. The conditions at Waikiki make are excellent for surfing.

415. According to the author, why is Waikiki ideal for surfing?
  a. The weather is great and the water is warm.
  b. The waves break abruptly as they approach the shore.
  c. The waves at Waikiki are a communicated agitation.
  d. Waikiki has some of the biggest waves in the world.
  e. The waves break gradually as they approach the shore.

416. The word *shoals* in line 9 refers to
  a. the sand kicked up as the waves break upon the beach.
  b. water becoming shallower as it approaches the shore.
  c. the steep cresting of a wave.
  d. the salty smell of the sea.
  e. water becoming deeper as you move away from the shore.

417. What part of a wave is responsible for the forming of surf?
  a. the upper portion of the wave
  b. the lower portion of the wave
  c. the strongest part of the wave
  d. the trailing portion of the wave
  e. the roaring part of the wave.

418. The word *impetus* in line 40 most nearly means
  a. a moving force.
  b. a serious obstacle.
  c. a slight annoyance.
  d. a slight hindrance.
  e. an area of very warm water.

419. The author's description of the transformation of a smooth undulating wave to a breaking wave (lines 18–21) indicates that
  a. The distance of a wave's break is dependent upon the bottom of the approaching the shoreline.
  b. It is rare for a wave to break gradually.
  c. It common for a wave to break abruptly.
  d. The size of a wave has to do with its speed through the water.
  e. A wave only travels through deep water.

420. The sentence *A wave is a communicated agitation* (line 1) is best
defined by which statement?
   a. the roar of a wave sounds angry when it breaks upon the shore.
   b. waves are a display of the ocean's fury.
   c. a wave is a surging movement that travels through the water.
   d. the size of a wave can vary.
   e. the ocean has baffled sailors for centuries.

421. What is the *secret* referred to in line 35?
   a. why a good wave for surfing must to be at least six feet tall
   b. A six-foot wave is between a quarter mile and a half mile in length.
   c. how a surfer can slide down a six-foot wave for a quarter of mile
   d. The smarter surfers paddle out to the deep water to catch the
   best waves.
   e. The water that composes a wave remains with the wave until it
   reaches the shore.

Questions 422–430 are based on the following passage.

*This passage details the life and career of Althea Gibson, an African-American
pioneer in the sport of tennis.*

(1)   Today, watching Venus and Serena Williams dominate the sport of
women's tennis with their talent and flair, it is hard to imagine that just
over fifty years ago African-American tennis players were barred from
competing on the grandest stages of their sport. Jackie Robinson broke
(5)   the color barrier in Major League Baseball in 1947, but the walls that kept
African-Americans from playing professional sports did not come tumbling down overnight. Almost four years passed from Jackie Robinson's
major league debut until a female African-American made a similar
impact upon the sport of women's tennis. That woman's name was Althea
(10)   Gibson.
   Althea Gibson was born on a cotton farm on August 25, 1927, in Silver, South Carolina. The early stages of the Great Depression forced her
sharecropper father to move the family from the bucolic Silver to the
urban bustle of New York City when she was just three years old. As a
(15)   child growing up in the Harlem section of the Manhattan, Althea found
she had an affinity for athletics. Basketball and paddle tennis were her
favorite sports, and she excelled at both. In fact, her talent at paddle tennis was so remarkable that in 1939 she won her age group at the New
York City paddle tennis championships. Shortly after, a very good friend
(20)   of Althea's suggested that she try lawn tennis. She showed an incredible
aptitude for the sport and her play caught the attention of members of the

predominately African-American Harlem Cosmopolitan Tennis Club,
who helped her raise money to become a member. At the age of fourteen,
Althea took her first real tennis lesson at the club under the tutelage of
(25) one-armed tennis coach Fred Johnson. She would never look back.

A year later in 1942, the major governing body for African-American
tennis tournaments—the American Tennis Association (ATA)—sponsored
the New York Girls Singles Championship at Althea's club. With her
aggressive and dominating style of play, she won the title easily. It was her
(30) first of what was to be many victories, on and off the court.

Althea dropped out of high school shortly after winning the New York
Girls Championship. She found the classes boring and wanted to con-
centrate on tennis. Her decision raised many eyebrows amongst members
of the ATA, who had hoped that she would become one of the sport's new
(35) stars. She was encouraged to leave New York City and move to Wilmington, North Carolina to live with the family of Hubert Eaton, a wealthy
doctor who was active in the African-American tennis community. Dr.
Eaton welcomed Althea into his family. He not only offered her guidance
with her tennis career, he also convinced her to finish the remaining three
(40) years of high school. While living with the Eaton family in Wilmington,
she would travel around the country to compete in ATA tournaments. By
the time she graduated in 1949, Althea had already won the first two of
what would be ten consecutive ATA national titles. She was regarded by
many as one of the most impressive young talents in the female game,
(45) but because of segregation she was not permitted to practice on any of
the public courts in Wilmington. She was also yet to be invited to any
of the major segregated tournaments.

By early 1950 Althea was making some headway. She was the first
African American to play in the national indoor tournament, where she
(50) finished second. Althea believed her two national championships and her
strong showing at the indoor tournament was proof that she was one of
female tennis's elite players. She and the ATA tried to lobby the United
States Lawn Tennis Association (USLTA) for an invitation to the 1950
U.S. Nationals, but despite the ATA's efforts and Althea's obvious merit,
(55) the USLTA failed to extend her an invitation.

Not every member of the USLTA was pleased with the organization's
decision. Former U.S. National and Wimbledon champion Alice Marble
wrote a scathing editorial in the July 1950 issue of *American Lawn Tennis*
magazine criticizing the USLTA's segregationist stance. Ms. Marble
(60) wrote, "The entrance of (African-Americans) into national tennis is as
inevitable as it has proven in baseball, in football, or in boxing; there is no
denying so much talent. . . . If Althea Gibson represents a challenge to the
present crop of players, then it's only fair that they meet this challenge on

(65) the courts." The editorial caused a national uproar that quickly led the USLTA to finally extend Althea an invitation to play in the 1950 U.S. Nationals tournament. This invitation would open many doors for Althea, and the following year she was the first African American to compete at Wimbledon.

(70) It took a few years for Althea to adjust to the world-class level of play. She won her first major tournament in 1956 and would dominate the sport for the next five years, winning six doubles titles and a total of eleven Grand Slam events including the U.S. Nationals and Wimbledon twice. Yet even at the height of her career as an international tennis champ, Althea was forced to endure discrimination. She was often refused hotels

(75) rooms and reservations at restaurants simply because of her skin color.

Althea once said that her extraordinary success was the product of being "game enough to take a lot of punishment along the way." The pioneering example set by Althea Gibson paved the way for future generations of African-American tennis players, and proved that beyond her

(80) tennis glory she was a true champion of the human spirit.

422. What is the main purpose of the passage?
   a. to glimpse a piece of the past
   b. to glorify athletes
   c. to disparage segregation
   e **d.** to teach the history of tennis
   e. to tell a story

423. The word *bucolic* in line 13 most nearly means
   a. rural.
   b. urban.
   c. sickly.
   a **d.** depressing.
   e. wealthy.

424. All of the following questions can be answered based on information from the passage EXCEPT
   a. What factors influenced the USLTA to invite Althea Gibson to the U.S. Nationals?
   e **b.** Did Althea play in another ATA tournament after she was invited to the U.S. Nationals?
   c. Why did Althea go to live with Dr. Eaton?
   d. To what specific types of discrimination was Althea subjected?
   e. How many times did Althea compete at Wimbledon?

425. Which of the following best describes the USLTA's change of heart regarding Althea's invitation?
   a. buckling under the pressure of public opinion
   b. a calculated strike against segregation
   c. a sudden recognition of Althea's abilities
   d. a bold marketing strategy
   e. a desire to diversify the women's game

426. The author uses Althea's quote about being *game enough* in line 80 to illustrate that
   a. Althea's career was plagued with injuries.
   b. the sport of tennis is more grueling than people realize.
   c. Althea believed the discrimination she faced served only to make her a stronger competitor.
   d. Althea was often fined for yelling at the referee.
   e. Althea believed talent was more important than mental toughness.

427. Althea's achievements are best described as
   a. remarkable displays of talent and athleticism.
   b. groundbreaking triumphs in the face of adversity.
   c. important events that led to immediate civil rights reform.
   d. one woman's fight against the world.
   e. historically insignificant.

428. Which statement best summarizes Alice Marble's quote in lines 60–64?
   a. Baseball, football, and boxing are more entertaining than tennis.
   b. Talent should dictate who could be a champion at a USLTA tournament, not race.
   c. There are players in the U.S. Nationals who do not deserve to be there.
   d. The USLTA should do away with invitations and make the tournament open to anybody.
   e. The ATA and USTLA should merge for the benefit of the sport.

429. Why did Althea's friend suggest that she try lawn tennis?
   a. Lawn tennis is a more competitive game than paddle tennis.
   b. The friend preferred playing lawn tennis.
   c. There was more money to be made playing lawn tennis than paddle tennis.
   d. The friend thought Althea might enjoy playing lawn tennis, and excel at it.
   e. The friend was looking for a tennis partner.

430. All of the following statements are supported by the passage EXCEPT
   a. Alice Marble was a white tennis player.
   b. Dr. Eaton's guidance helped Althea's career.
   c. Althea won the New York Girls Singles championship when she fifteen.
   d. The public tennis courts in Wilmington were segregated.
   e. Althea Gibson won more Grand Slam titles than any other female tennis player.

Questions 431–439 are based on the following passage.

The following passage chronicles the 1919 "Black Sox" baseball scandal.

(1)   Professional baseball suffered during the two years the United States was involved in World War I. Many Americans who were preoccupied with the seriousness of the war raging overseas had little concern for the trivialities of a baseball game. After the war ended in 1919, many
(5)   Americans wanted to put those dark years behind them and get back to the normal activities of a peaceful life. One of those activities was watching baseball. In the summer of 1919, ballparks that just one year earlier had been practically empty were now filled daily with the sights and sounds of America's favorite pastime. That year, both the Cleve-
(10)   land Indians and New York Yankees were two of the strongest teams in baseball's American League, but one team stood head and shoulders above the rest: The Chicago White Sox.

   The Chicago White Sox, called The White Stockings until 1902, were owned by an ex-ballplayer named Charles Comiskey. Between the
(15)   years of 1900 and 1915 the White Sox had won the World Series only once, and Comiskey was determined to change that. In 1915, he purchased the contracts of three of the most promising stars in the league: outfielders "Shoeless" Joe Jackson and "Happy" Oscar Felsch, and second baseman Eddie Collins. Comiskey had only to wait two years for his
(20)   plan to come to fruition; the 1917 White Sox, playing in a park named

for their owner, won the World Series. Two years later they had the best record in all of baseball and were again on their way to the Series.

(25)

(30)

(35)

Baseball players' salaries in that era were much different than the exorbitant paychecks of today's professional athletes. Often, ballplayers would have second careers in the off-season because of the mediocrity of their pay. To make matters worse, war-torn 1918 was such a horrible year for baseball attendance that many owners cut player salaries for the following season. However, it is said in all of baseball there was no owner as parsimonious as Charles Comiskey. In 1917 he reportedly promised every player on the White Sox a bonus if they won the American League Championship. After winning the championship, they returned to the clubhouse to receive their bonus—a bottle of inexpensive champagne. Unlike other owners, Comiskey also required the players to pay for the cleaning of their uniforms. The Sox had the best record in baseball, but they were the least paid, were the most discontented, and wore the dirtiest uniforms.

(40)

Comiskey's frugality did not sit well with the players. They were most upset with the fact that he did not raise salaries back to their 1918 levels, even though the ballpark attendance figures for 1919 were higher than any previous year. One player, Eddie Ciccotte, felt especially ill-treated by Comiskey. The owner promised the pitcher a bonus of $10,000 if he won thirty games, but after Ciccotte won his twenty-ninth game he was benched by Comiskey for the rest of the season.

(45)

(50)

Gamblers were such a common sight around the Chicago ballpark that Charles Comiskey had signs proclaiming "No Betting Allowed In This Park" posted conspicuously in the stands. The money with which these gamblers tempted the players was hard to refuse, and it was rumored that to supplement their income some of the lower-paid athletes would offer inside tips to the bettors. But gamblers' mingling with ballplayers wasn't solely confined to the White Sox. In 1920, allegations involving gambling among Chicago Cubs players brought to light a scandal that would shock Chicago and the rest of America: Eight members of the White Sox had thrown the 1919 World Series.

(55)

(60)

The exact facts regarding the scandal will never be known, but the most accepted theory is that just prior to the World Series, White Sox player Chick Gandil had approached a gambler by the name of Joseph Sullivan with a proposal that for $100,000 Gandil would make sure the Sox lost the Series. Gandil needed to recruit other players for the plan to work. It was not hard for him to do—there were many underpaid players on the White Sox who were dissatisfied with the way Comiskey operated the team. Ultimately, the seven other play-

ers that were allegedly involved in the scheme were Eddie Cicotte,
Happy Felsch, Joe Jackson, Fred McMullin, Charles "Swede" Ris-
(65) berg, Buck Weaver, and Claude Williams.

They were successful. The Chicago White Sox, heavily favored to
beat an inferior Cincinnati Reds team, lost the nine-game World Series
in eight games, due in most part to the inferior play of the eight con-
spiring players. When the scandal made headlines the following year
(70) the press began to refer to them as the Black Sox, and the ignominious
label would be used to describe them forever.

When the eight players stood before an Illinois grand jury, it was
determined that that there was not enough substantial evidence for
any convictions, and the players were all eventually acquitted of any
(75) criminal wrongdoing. Interestingly enough, Charles Comiskey paid
for the players' high-priced defense lawyers. Unfortunately for
Comiskey, there was to be no similar reprieve from major league baseball: Every single one of the accused players was banned from the
game for life. Comiskey's once mighty team was decimated by the loss
(80) of its most talented players, and the 1921 White Sox finished the season in seventh place.

431. According to the passage, who was the supposed ringleader of the
Black Sox scandal?
  a. Charles Comiskey
  b. "Shoeless" Joe Jackson
  c. Eddie Ciccotte
  d. Eddie Collins
  e. Chick Gandil

432. In line 29, the word *parsimonious* most nearly means
  a. generous.
  b. stingy.
  c. powerful.
  d. friendly.
  e. jovial.

433. According to facts from the passage, what was the name of the
White Sox's ballpark?
  a. Chicago Park
  b. Comiskey Park
  c. Sullivan Stadium
  d. White Sox Park
  e. Sox Field

434. In line 54, the word *thrown* refers to
  a. losing intentionally.
  b. pitching a baseball.
  c. projecting upon.
  d. dashing upon.
  e. abandoning something.

435. According to the passage, how many World Series' did the White Sox win between 1900 and 1919?
  a. none
  b. one
  c. two
  d. three
  e. four

436. All of the following questions can be answered based on information from the passage EXCEPT
  a. Who was the second baseman for the 1915 White Sox?
  b. Did the White Sox play in the American League or the National League?
  c. What was the White Sox's original name?
  d. How many games did Eddie Ciccotte pitch in 1918?
  e. Why did many baseball owners lower player salaries for the 1919 season?

437. In lines 71–72, word *ignominious* most nearly means
  a. uneducated.
  b. dishonorable.
  c. exalted.
  d. worthy.
  e. unentertaining.

438. The last paragraph of the passage suggests that Charles Comiskey
  a. thought the team was better off without the eight players.
  b. hoped all eight players would be convicted and sent to jail.
  c. wanted the players involved in the scandal to return to the team.
  d. was contemplating retirement.
  e. had a plan to get the White Sox back to the World Series.

439. The passage as a whole suggests that
   a. The White Sox probably fixed the 1917 World Series, too.
   b. Charles Comiskey may have been in part to blame for his play-ers' actions.
   c. ballplayers betting on games was a highly unusual occurrence.
   d. baseball never recovered after World War I.
   e. Charles Comiskey often bet against his own team.

5-2-14

2-16-14 (-3)

11/10/18

4/12/19

Questions 440–449 are based on the following passage.

The following passage is adapted from a magazine article entitled **The Revival of the Olympic Games: Restoring the Stadium at Athens,** *published prior to the first modern Olympics.*

(1)   For several months an unwonted activity has prevailed in one quarter
      of Athens. Herodes Atticus Street behind the royal garden, one of the
      most retired streets of the city, has resounded all day long with the rat-
      tle of heavy wagons bringing blocks of marble from Pentelikon. At
(5)   sunrise and sunset crowds of workingmen are seen moving through
      this street, the lower end of which opens upon a bridge across the Ilis-
      sos, and on the opposite bank lies the Panathenaic Stadium, now being
      lined with marble for the Olympic games which are to be held in it
      early in April. The time is short, and the work is being pressed for-
(10) ward. When the International Athletic Committee, at a session in
      Paris last year, decided to have a series of athletic contests once in four
      years in various countries, it is not surprising that they selected Greece
      for the first contest. Although Greece now has as little of the athletic
      habit as any nation of the civilized world, its past is interwoven with
(15) athletics. Olympia is a magic word, and the committee were doubtless
      swayed partly by sentimental reasons in the choice of name and place.
        But some may wonder why, since the games come to Greece, they
      are not to be held at Olympia, to justify the name which they have
      taken. This is because the originators of the scheme, although they
(20) have conceded something to sentiment, are no visionaries, but men of
      practical common sense. Even their concession to sentiment is likely
      to turn out to be a clever piece of practical management, calculated to
      launch the games upon the world with more success than could have
      been secured in any other way. The games also have a name which will
(25) be just as true in 1900 at Paris, and 1904 in America, as it is this year
      in Athens.
        Now, however fine a thing it might be to let athletes stir real
      Olympic dust, and to let runners put their heels into the very groove

of the old starting-sill, with the feeling that thirty centuries looked
(30) down upon them, it would not be practical. A successful athletic contest cannot be held in the wilderness. It demands a crowd and sustenance for a crowd. The crowd is the one essential concomitant of the
athletes. But a crowd will not go where it cannot eat and sleep. To
bring to Olympia a concourse sufficient to in modern times make the
(35) games anything like a success would demand the organization of a
first-class commissary department, and that too for a service of half a
month only. Shelter and food for such an occasion come naturally only
in connection with some city with a market. Ancient Olympia, with all
its magnificent buildings, was of course that sort of city, albeit practi-
(40) cally a deserted city except for a few days once in four years.

The visitors at Athens next April—and it is hoped that there will be
tens of thousands of them—will doubtless feel keenly enough the
inadequacy even of a city of 130,000 inhabitants, to give them all that
they seek in the way of material comforts. The problem of seating a
(45) large crowd of spectators did not come up before the International
Committee. But it is this problem which has found a most happy solution in Athens. The Stadium at Olympia, although excavated at each
end by the Germans, still lies in most of its course under fifteen or
twenty feet of earth. But the Stadium at Athens has always been a fit
(50) place for a monster meeting, provided people would be contented to
sit on its sloping sides without seats. When a local Athenian committee was formed, composed of most of the citizens conspicuous for
wealth or position, and some resident foreigners, under the presidency
of Constantine, crown prince of Greece, one of the first questions
(55) before it was this question of seating; and its attention was naturally
directed to the Stadium.

A wealthy and generous Greek of Alexandria, George Averoff, who
was known as a man always on the watch to do something for Athens,
readily took upon himself the expense of restoring the Stadium to
(60) something like its former splendor, when it was lined with marble and
seated fifty-thousand spectators. He has already given over nine hundred thousand drachmas, which, if the drachma were at par, would be
$180,000, but which now amounts to only about $100,000. There is
a sub-committee of the general committee above described, desig-
(65) nated as the committee on the preparation of the Stadium, composed
of several practical architects, but including also the Ephor General of
Antiquities, and the directors of the foreign archaeological schools.
The presence of the archaeological element on this committee emphasizes the fact that the new work is to be a restoration of the old.

440. In line 1, the word *unwonted* most nearly means
   a. not welcome.
   b. out of the ordinary.
   c. unexpected.
   d. ancient.
   e. nocturnal.

441. Herodes Atticus Street (line 2) is located where in relation to the Stadium at Athens?
   a. behind the royal garden
   b. on Mount Olympus
   c. across the Illissos river
   d. just north of Pentelikon
   e. directly adjacent to

442. Based on information in the passage, what year were the first modern Olympics to be held?
   a. 1892
   b. 1896
   c. 1900
   d. 1904
   e. 1908

443. One of the *sentimental reasons* the author refers to in line 16 is
   a. Athens was always the largest city in Greece.
   b. Panathenaic Stadium is the oldest stadium in Ancient Olympia.
   c. Olympia, Greece was the site of the original Olympics.
   d. Paris was a better choice for the first modern Olympic games.
   e. George Averoff was once the King of Greece.

444. All of the following are reasons why the first modern games were held in Athens and not in Olympia EXCEPT
   a. Olympia was a much smaller city than Athens.
   b. Parts of the Stadium at Olympia were buried underground.
   c. Athens offered better facilities for the crowd in terms of food and shelter.
   d. The Germans voted against Olympia in favor of Athens.
   e. The city of Olympia would not attract the same crowd as Athens.

445. Who was in charge of solving the problem of seating the crowds
expected at Athens?
   a. the International Athletic Committee
   b. the Germans
   c. George Averoff
   d. the Ephor General of Antiquities
   e. a local Athenian Committee

446. According to the passage, about how long were the games to be?
   a. two weeks
   b. the month of April
   c. four years
   d. three weeks
   e. a few days

447. In line 62, the word *drachma* refers to
   a. a block of marble.
   b. the Greek word for marble.
   c. the name of Greek money.
   d. a type of stadium seat.
   e. a type of Greek food.

448. In line 30, what does the author claim *would not be practical*?
   a. trying to revive the spirit of the ancient games
   b. holding the new Olympics in Olympia
   c. excavating the Stadium at Olympia for use at the modern games
   d. refurbishing the Stadium at Athens
   e. seating fifty-thousand spectators

449. The phrase *the feeling that thirty centuries looked down upon them*
(lines 29–30) refers to the
   a. political importance of holding the first modern games at the
   site of Ancient Olympia.
   b. decision to hold the second modern Olympics in France.
   c. importance of reviving the spirit of the ancient Olympic games.
   d. sentimental value of holding the modern games at the site of
   Ancient Olympia.
   e. need for the best amateur athletes to compete.

Questions 450–460 are based on the following passages.

The following passages detail two very different perspectives of life aboard a ship in the age of sail. The first passage describes an English pleasure yacht in the early 1800s. The second passage recounts a young boy's impressions of the first time he set sail in a merchant vessel.

PASSAGE 1

(1)   Reader, have you ever been at Plymouth? If you have, your eye must have dwelt with ecstasy upon the beautiful property of the Earl of Mount Edgcumbe: if you have not been at Plymouth, the sooner that you go there the better. You will see ships building and ships in ordi-
(5)   nary; and ships repairing and ships fitting; and hulks and convict ships, and the guard-ship; ships ready to sail and ships under sail; besides lighters, men-of-war's boats, dockyard-boats, bum-boats, and shore-boats. In short, there is a great deal to see at Plymouth besides the sea itself: but what I particularly wish now is, that you will stand at the bat-
(10)   tery of Mount Edgcumbe and look into Barn Pool below you, and there you will see, lying at single anchor, a cutter; and you may also see, by her pendant and ensign, that she is a yacht.

        You observe that this yacht is cutter-rigged, and that she sits grace-fully on the smooth water. She is just heaving up her anchor; her fore-
(15)   sail is loose, all ready to cast her—in a few minutes she will be under way. You see that there are ladies sitting at the taffrail; and there are five haunches of venison hanging over the stern. Of all amusements, give me yachting. But we must go on board. The deck, you observe, is of narrow deal planks as white as snow; the guns are of polished
(20)   brass; the bitts and binnacles of mahogany: she is painted with taste; and all the moldings are gilded. There is nothing wanting; and yet how clear and unencumbered are her decks! Let us go below.

        There is the ladies' cabin: can anything be more tasteful or elegant? Is it not luxurious? And, although so small, does not its very confined
(25)   space astonish you, when you view so many comforts so beautifully arranged? This is the dining-room, and where the gentlemen repair. And just peep into their state-rooms and bed-places. Here is the stew-ard's room and the buffet: the steward is squeezing lemons for the punch, and there is the champagne in ice; and by the side of the pail
(30)   the long-corks are ranged up, all ready. Now, let us go forwards: here are, the men's berths, not confined as in a man-of-war. No! Luxury starts from abaft, and is not wholly lost, even at the fore-peak. This is the kitchen; is it not admirably arranged? And how delightful are the

(35) fumes of the turtle-soup! At sea we do meet with rough weather at
times; but, for roughing it out, give me a *yacht*.

PASSAGE 2

(1) My very first sea voyage was in a small merchant vessel out of New
York called the *Alba*. I was only twelve years old at the time, and full
of dreams of boundless adventure upon the high seas. I was to serve
as the ship's boy. I was given the post by my Uncle Joseph, the weath-
(5) ered old captain of the *Alba* who uttered few words, choosing to speak
more with his menacing gaze than with his mouth. The moment I
stepped upon the bustling deck my Uncle Joseph set me straight about
shipboard life. There were to be no special privileges afforded to me
because of our relations. I was to live and mess in the 'tween decks
(10) with the other seamen, and because I was his nephew, I would probably have to work twice as hard as the others to prove my worth. From
that point on I was to refer to my uncle as "Sir" or "Captain," and only
speak to him when he addressed me. He then told me a bit about the
Alba. I learned that she was a cutter, and all cutters were fore-and-aft
(15) rigged, and possessed only a single mast. After my brief lesson, he then
sent me below deck to get myself situated.

What I found when I dismounted the ladder below was an entirely
different world than the orderly brightness of the top deck. Here was
a stuffy and dimly lit space barely tall enough for me to stand up
(20) straight in. It was the middle of July, and the heat was oppressive.
There seemed to be no air at all, there certainly were no windows, and
the stench that rose up from the bilge was so pungent it made me gag.
From the shadows, a pair of eyes materialized. They belonged to a
grimy boy no older than me.
(25) "Hello mate, you must be the new lubber just shipped aboard. I'm
Nigel. Follow me, we're just in time for dinner."

My new friend led me into the tiny dining room where the crew
messed. The men ate shoulder to shoulder on wooden tables bolted to
the deck. The horrific smell of so many men crammed together was
(30) overpowering. We received our food from the ship's cook, a portly
man in a filthy apron who, with the dirtiest hands I'd ever seen, ladled
us out a sort of stew. We found two open spots at a mess table and sat
down to eat. The stew was lukewarm and the mysterious meat in it was
so tough I could barely chew it. I managed to swallow a few spoonfuls
(35) and pushed my dish aside.

With a smile that was graveyard of yellow sincerity, Nigel pushed the dish back to me and said, "I'd get used to the grub, mate. It ain't so bad. Besides, this is the freshest it'll be on the voyage."

After dinner, Nigel showed me our berth. It was a tiny lightless cub-
(40) byhole near the bow of the boat that was barely six feet long and only five feet high. There was a small area where I could stow my clothes, and at night we would string up our hammocks side by side with two other boys, both of whom were on duty at the moment.

That night when we were under way, the boat ran into a vicious
(45) Atlantic storm. The waves tossed the *Alba* around like it was a tiny raft. The ship made such noises; I was afraid it would simply break apart at any moment. The seawater that crashed upon the deck leaked through the planks and dripped upon my head. It would have bothered me if I were not already horribly seasick. As I lay there miserably rocking
(50) back and forth in my damp hammock, I asked myself, "What have I gotten myself into?"

450. According to both passages, it is not uncommon for ships to
   a. meet rough seas.
   b. run out of fresh drinking water.
   c. not return home for quite a while.
   d. leak in heavy weather.
   e. have children onboard.

451. In the last sentence of Passage 2 the narrator suggests that he
   a. may never recover from the seasickness.
   b. does not like Nigel.
   c. made a mistake taking the voyage aboard the *Alba*.
   d. should have eaten the stew.
   e. should have stayed in school.

452. Which statement best summarizes the narrator's description of Plymouth in lines 3–8?
   a. The port at Plymouth is full of rowdy sailors.
   b. Plymouth is a dreary and overcrowded place.
   c. Plymouth is a deserted and over-industrialized area
   d. There are many interest sights to behold at Plymouth.
   e. The British Royal Navy anchors at Plymouth.

453. What do the yacht in Passage 1 and the *Alba* in Passage 2 have in common?

   a. They were both built in England.
   b. They both have only a single mast.
   c. They are both made of iron.
   d. They both have lifeboats.
   e. They are both fast.

454. How do the yacht in Passage 1, and the *Alba* in Passage 2 differ?

   a. The yacht does not carry cargo.
   b. The yacht is much bigger than the *Alba*.
   c. There are no passengers aboard the Alba, only crew.
   d. The yacht is much more luxurious than the *Alba*.
   e. The yacht is much faster than the *Alba*.

455. Why does the captain in Passage 2 (lines 11–12) demand that his nephew call him *Sir* or *Captain*?

   a. The captain wanted his nephew to understand who was in charge.
   b. The captain did not want any member of the crew to know the narrator was his nephew.
   c. The captain was afraid that if he showed affection to his nephew, he would lose his authority over the crew.
   d. The captain was not really the narrator's uncle.
   e. It was important that the crew understood that the boy was no more privileged than anyone else aboard.

456. In Passage 1, line 26, the use of the word *repair* most nearly means

   a. go.
   b. fix things.
   c. sit in pairs.
   d. get dressed.
   e. exercise.

457. The narrator of Passage 1 most probably

   a. is a seasoned sea captain.
   b. is very wealthy.
   c. is an experienced yachtsman.
   d. suffers from seasickness.
   e. was in the Royal Navy.

458. In Passage 2, line 36, the narrator describes Nigel's smile as *a graveyard of yellow sincerity*. What figure of speech is the narrator employing?
  a. onomatopoeia
  b. simile
  c. personification
  d. alliteration
  e. metaphor

459. Together, these two passages illustrate the idea that
  a. the reality of two seemingly similar situations can often be extremely different.
  b. boating is a very dangerous pastime.
  c. dreams sometimes fall very short of reality.
  d. Plymouth is much nicer than New York.
  e. hard work pays off in the end.

460. The word *berth*, found in Passage 1, line 31 and Passage 2, line 39 most nearly means
  a. a sailor's hometown.
  b. the sleeping quarters aboard a boat.
  c. the kitchen aboard a boat.
  d. the bathroom aboard a boat.
  e. the lower deck of a boat.

Answers

400. b. The author's tone in this passage could only be described as *light-hearted*. The subject of the passage itself is not of a particularly serious nature, and the author's deduction in lines 21–23 that watching a sport on television would technically characterize couch potatoes as athletes is humorous and subtly mocks those who would argue over what is a "true" sport.

401. e. *Vigorous*, as it is used in the passage, is an adjective that describes an activity carried out forcefully or energetically. In other words, a vigorous activity requires a *physical exertion* (line 11) that would cause one to *break a sweat* (line 12). This type of activity is best described as strenuous, choice **e**.

402. d. *Cheerleading* (choice **a**), *skiing* (choice **b**), *race car driving* (choice **c**), and *gymnastics* (choice **e**) are all strenuous activities that require good hand-eye coordination and run the risk of injury.

Playing *horse shoes* (choice **d**) only requires good hand-eye coordination.

403. **b.** *Bellicose* most closely means warlike. There are two major clues in this passage to help you answer this question. The first clue lies in the translation of the name *Tewaarathon,* meaning "Little Brother of War." Another clue lies in lines 18–19, where the passage states that these games were *excellent battle preparation for warriors.*

404. **c.** The answer to this question can be found in lines 17–20, as well as in the entire second paragraph. The passage states that the games played by the Native Americans were often substitutes for war, and from time to time the games held religious and spiritual significance. Don't be fooled by choice **e**; the Native Americans may have played friendly exhibition matches, but this is not discussed anywhere in the passage.

405. **a.** "Little Brother of War" is the best choice for the title of this passage because, in the first paragraph, the games are described as fierce and warlike. Choice **a** is also the name of the original Iroquois game, which was the subject of the entire second paragraph. The other choices do not fit because they are unsupported by the passage, or describe only a small portion of the passage.

406. **c.** The answer can be found in the two sentences that follow the phrase. The sentences state that the games were often high-stakes substitutes for war, and it was not uncommon for players to suffer serious injuries at the hands (and sticks) of others. These statements describe the fierce nature of the games, and suggest that players would not hesitate to resort to violent tactics to score, *by any means necessary.* Choices **d** and **e** are true and mentioned in the passage, but they do not fit in context with the phrase.

407. **e.** The author's primary purpose in writing this passage is to illustrate the importance of these games in Native American culture. The author does this by giving examples of the spiritual and peacekeeping significance of the games to the Native Americans. The passage does inform us that lacrosse evolved from these ancient games, but it does not specifically describe any aspect of modern lacrosse or any other sport, therefore choices **a** and **c** are incorrect. Choices **b** and **d** are both mentioned by the author, but they are not the main subjects of the passage, and nowhere in the passage does the author condone or condemn the violence of the games.

408. **e.** The author's primary purpose in writing this passage is to discuss his belief that commercialism's strong presence in today's society strongly influences a person's view of his or her personal identity. A good illustration of this can be found on line 23–24, where the author states, *we are prepared to spend our way into a trendy identity*.

409. **b.** The *commercial range of options* in line 3 is the numerous products available for purchase by today's consumer. Line 6 holds a clue to answering this question: The author refers to the modern practice of wearing old symbols such as a kilt as *the personal choice of a particular consumer*.

410. **b.** The term *disposable income* refers to the specific amount of a person's income that is allotted as spending money. This is the only choice that makes sense in the context of the passage.

411. **d.** The statement that one *can tell a lot about somebody by what they are wearing* is directly contradicted by the claim the author makes in lines 3–6: *Seeing someone in a Harvard or Oxford sweatshirt or a kilt or a military tie now communicates nothing at all significant about that person's life other than the personal choice of a particular consumer*.

412. **a.** The author's point of view of today's society in lines 12–14 is that today's world is much smaller and more hectic than it used be, which makes it harder for people to put down solid roots and identify with a singular way of life. In short, *times have changed*.

413. **c.** In line 27 the author states a surfer should lie upon a surfboard *like a small boy on a coaster*, and then goes on in lines 32–33 to say that the surfer slides down a wave *just as a boy slides down a hill on his coaster*.

414. **d.** The question asks for the statement that cannot be answered based on information given in the passage. In lines 25–29, the author describes the shape and dimensions of a flat board, and tells the reader how to paddle and lie upon it. But nowhere in the passage does the author state that a flat board is the most popular type of surfboard.

415. **e.** The answer to this question is found in lines 18–22. The author states that *the bottom shoals gradually from a quarter of a mile to a mile* toward the beach at Waikiki, producing *a splendid surf-riding surf*.

416. **b.** When the word *shoal* is used as a verb it usually means to become shallow (as in water) or to come to a shallow or less deep part of. Lines 9–10 state that as the wave approaches the shore *the lower portion of the wave strikes land first and is stopped.* If the sea bottom is rising, the water will therefore be not as deep, in other words—it will be *shallower*.

417. **b.** The answer is explained in lines 9–17, and spelled out in lines 16–17: *It is the bottom of a wave striking against the top of the land that is the cause of all surfs.*

418. **a.** As it is used in the passage, *impetus* most nearly means *a moving force*. In this case, a wave is a moving force through the water. If you did not know the correct definition, the best way to answer this question would be to replace *impetus* in the sentence with each of the given answer choices to see which one makes the most sense in context.

419. **a.** The best approach to this question is to reread lines 18–21 for each answer choice to see which choice is directly supported by the given text. For this question you would not have to go far to find the answer: choice **a** quickly summarizes the text of those lines. All the other answer choices are unsupported or contradicted by the given text.

420. **c.** Context clues are your best aid in answering this question, and an important context clue is given in lines 1 and 2. The author goes on to state that the water that composes *the body of a wave is stationary*, and gives the example of the thrown stone causing ripples in the water. The rock that is thrown is the cause of the agitation of the water. The ripples (or the waves) that surge away from that agitation are the communication of that agitation moving through the water. Therefore, choice **c** is the correct answer.

421. **c.** In line 33, the author compares surfing to *slid[ing] down a hill*. But unlike a six-foot hill, a surfer can slide down a six-foot wave for more than a quarter of a mile without ever reaching the bottom. The author explains that this is possible because the water that composes the wave is, like a hill, standing still *and new water is rising into the wave as fast as the wave travels*, preventing the surfer from reaching the bottom (lines 41–43). So while it looks like a surfer is sliding along moving water, he or she is actually stationary on a wave as it moves through the water. That's the *secret*.

422. **e.** *Glimpsing a piece of the past* (choice **a**), *glorifying athletes* (choice **b**), *disparaging segregation* (choice **c**), and learning some *tennis history* (choice **d**) are all story elements that support the main purpose of the passage: To tell the story of Althea Gibson, the woman who broke the color barrier in professional tennis (choice **e**).

423. **a.** The word *bucolic* is most often used to describe something typical of or relating to rural life. If you did not know what bucolic meant, there are contextual clues to help you. In lines 11-15, the passage tells us that Althea was born on a *cotton farm* and her father was a *sharecropper*. Also, in lines 13–14, the author contrasts the *bucolic Silver* with New York City's *urban bustle*.

424. **e.** The passage states that Althea Gibson was a two-time Wimbledon champion. However, the passage does not offer the exact number of defeats Althea suffered at Wimbledon in her career.

425. **a.** Althea's accomplishments in 1949 and 1950 should have earned her an invitation to the 1950 U.S. Nationals, but her and the ATA's efforts to secure an invitation from the USTLA fell on deaf ears (lines 51–57). It was not until the national uproar spurred by Alice Marble's editorial (lines 62–66) that the USTLA, buckling under the weight of public pressure (choice **a**), relented and extended Althea an invitation to play.

426. **c.** Althea was an extraordinarily gifted athlete, yet because of the color of her skin and the time in which she lived, her path to success from the very beginning was obstructed by segregation and discrimination. Althea was not allowed to practice on public tennis courts (lines 47–48), barred from USLTA-sponsored events (line 57), and was refused hotel rooms and restaurant reservations (lines 76–78). Althea's ability to put these distractions aside and excel was a triumph of mental toughness, and the author uses the quote on line 80 to illustrate that fact.

427. **b.** When looking at questions such as this one, it's important to think each choice through before hastily picking an answer. This question has two tough distracters: choices **c** and **d**. At first glance, choice **c** seems like a good pick, but the word *immediate* is what makes it incorrect. Althea Gibson's achievements were certainly victories for the civil rights movement, but in lines 6–7 it is stated that the color barrier *did not come tumbling down overnight*. Choice **d** is attractive, but Althea did not take on the world alone. The ATA and people like Dr. Eaton and Alice Marble all had a hand in guiding and assisting Althea on her pioneering path. Choice **e** is incorrect because Althea's historic achievements on and off the court were groundbreaking, and she accomplished it all in the face of adversity.

428. **b.** Alice Marble believed that talent should decide who can be a champion, not race (choice **b**). Nowhere in her comments did Alice Marble say baseball, football, and boxing are more entertaining than tennis (choice **a**), or that there were undeserving players in the U.S. Nationals (choice **c**). Nor did she propose that the USLTA make the tournament open to anybody (choice **d**).

429. **d.** Althea's friend probably suggested that Althea try lawn tennis because she was a champion paddle tennis player and enjoyed the sport very much (lines 16–17). The other choices either don't make sense or are not supported by facts from the passage.

430. **e.** In lines 71–75, the passage states that Althea won a total of eleven Grand Slam titles in her career. However, nowhere in the passage does it state that those eleven titles were a record number for a female.

431. **e.** The answer is found in line 58 of the passage. Chick Gandil first approached the gambler with his scheme, and then recruited the seven other players.

432. **b.** *Parsimonious* is a word used to describe someone who is frugal to the point of stinginess. Comiskey's pay cuts (line 27), bonus of cheap champagne (lines 32–33), refusal to launder uniforms (lines 33–34), and his benching of Eddie Ciccotte (lines 42–44) are all clues that should help you deduce the answer from the given choices.

433. **b.** Answering this question involves a bit of deductive reasoning. Though the actual name of the ballpark is never given in the passage, lines 20–21 state that the 1917 White Sox won the World Series *playing in a park named for their owner.*

434. **a.** As it is used in line 54, *thrown* means to have lost intentionally. The answer to this question is found in lines 59–60. For $100,000 Chick Gandil *would make sure the Sox lost the Series.*

435. **c.** Lines 14–16 state *between the years of 1900 and 1915 the White Sox had won the World Series only once,* and then line 21 tells us they won it again in 1917. Be careful not to mistakenly select choice **d**, three; the question asks for the number of World Series the Sox *won,* not the number of Series played.

436. **d.** In lines 42–44 the author states that *after Ciccotte won his twenty-ninth game he was benched by Comiskey for the rest of the season.* Choice **d** asks for the number of games he *pitched.* It is stated that he pitched and won twenty-nine games in 1919, but the passage doesn't mention the number of games he pitched in which he lost, so you can't know for sure.

437. **b.** *Ignominious* is a word used to describe something marked with shame or disgrace, something dishonorable. The *ignominious label* referred to in lines 71–72 is *Black Sox*—the nickname the Chicago press took to calling the scandalized and disgraced White Sox team.

438. **c.** It is stated throughout the passage Comiskey was a frugal man, yet in lines 76–77 it says that he paid for the players' defense lawyers. Why? The answer to that and the biggest clue to answering this question lies in the last sentence of the passage: *Comiskey's once mighty team was decimated by the loss of its most talented players, and the 1921 White Sox finished the season in seventh place.*

439. **b.** Lines 47–50 state that gamblers would often target with *the lower-paid athletes* because *the money with which these gamblers tempted the players was hard to refuse.* The passage tells that due to Charles Comiskey's stinginess with his players, *there were many underpaid players on the White Sox who were dissatisfied* (lines 61–62) and they *were the most discontented* team in baseball (line 35). These factors suggest that if Charles Comiskey had treated his players better, perhaps they might not have been so eager to betray him.

440. **b.** A context clue to help you answer this question is found in lines 2–3, when the author states that Herodes Atticus Street is *one of the most retired streets of the city.* Of the given answer choices, *out of the ordinary* best describes the activity of heavy construction on a normally quiet street.

441. **c.** The author states in lines 6–7 that the lower end of Herodes Atticus Street *opens upon a bridge across the Ilissos, and on the opposite bank lies the Panathenaic Stadium*—the Stadium at Athens.

442. **b.** Lines 11–12 state the Committee decided that the Olympics would be held *once in four years*, and the next two Olympics to follow would be held in the years 1900 and 1904 (line 25).

443. **c.** As stated in line 16, the organizers of the first modern Olympics were *swayed partly by sentimental reasons in the choice of name and place.* The ancient Olympics took its name from the city where it was held every four years: Olympia, in Greece. To honor those ancient games, the organizers named the modern games the Olympics and would play the inaugural contests in Greece.

444. **d.** The Germans were involved in excavating the ancient Stadium at Olympia (lines 47–48). Nowhere in the passage does it mention that there was a vote to decide between Olympia and Athens.

445. **e.** Lines 44–46 state that *the problem of seating a large crowd of spectators did not come up before the International Committee* (choice **a**). In fact, it was a local Athenian committee (choice **e**) *composed of most of the citizens conspicuous for wealth or position, and some resident foreigners* (lines 52–53) that were posed with the question of seating for the games in Athens.

446. **a.** Lines 35–37 state that if Olympia were to be considered a viable site for the modern Olympics, it would *demand the organization of a first-class commissary department, and that too for a service of half a month only.* Half a month is roughly *two weeks,* choice **a.** It is true that line 40 states that the games were just a few days (choice **e**) every four years, but that is in reference to the ancient Olympic games.

447. **c.** Before Greece switched to the Euro in 2002, Greek money was called *drachma.* The answer to this question lies in line 61–63, where it states that nine hundred thousand drachmas were worth about one hundred thousand dollars.

448. **b.** In lines 31–32 the author states that a successful athletic contest *cannot be held in the wilderness* and *demands a crowd and sustenance for a crowd.* Holding the games at Olympia would have sentimental value because of its history, but *it would not be practical* because Olympia does not have the proper facilities and resources to accommodate the crowds that would descend upon the games.

449. **d.** In lines 29–30, the author uses the phrase *the feeling that thirty centuries looked down upon them* to emphasize the sentimental value of holding the modern games at the site of Ancient Olympia (choice **d**). But the author goes on to say that despite the sentimental value, it just wouldn't be practical.

450. **a.** In lines 34–35, the narrator of Passage 1 mentions *At sea we do meet with rough weather at times.* In Passage 2, lines 44–45, the boy recounts that his boat *ran into a vicious Atlantic storm,* and *the waves tossed the* Alba *around like it was a tiny raft.* Choice **d** may seem like an attractive answer, but there is only evidence that the *Alba* leaks (line 47), not the yacht, and the question requires support from *both* passages.

451. **c.** In the last sentence of Passage 2 the narrator questions his decision to take the voyage aboard the *Alba* by asking himself *What have I gotten myself into?* This self-doubt indicates that he believed his decision may have been a mistake. This choice best answers the question.

452. **d.** In lines 2–3, the author of Passage 1 tells of the beautiful property belonging to the Earl of Mount Edgcumbe and implores the reader to visit Plymouth if they ever get the chance. He then goes on to describe the bustling harbor at Plymouth and finishes with: *there is a great deal to see at Plymouth besides the sea itself* (lines 8–9). In short, he describes all the interesting sights to behold at Plymouth. All the other choices either do not make sense or are not specifically supported by details from the text.

453. **b.** In lines 10–12 of Passage 1 the narrator states that the yacht is a particular type of ship known as a cutter. In lines 14–15 of Passage 2, the Captain explains to his nephew that the *Alba* is cutter, as well. In that same conversation the nephew learns that all cutters share a similar trait: they possess only *a single mast* (line 15). Therefore, choice **b** is the correct answer.

454. **d.** When answering this question, the key is to be sure to find the only choice that is supported by specific examples from the text. Nowhere in the text of Passage 1 does it state that the yacht carries cargo, but on the other hand it never mentions the fact that it does not. The same reasoning goes for choices **b**, **c**, and **e**. The yacht may be bigger and faster than the *Alba*, and the *Alba* may carry only crew, but these facts are never mentioned in the texts so we can't know for sure. That leaves only one possible answer: choice **d**. The yacht is most certainly more luxurious than the *Alba*, and this statement is backed by both narrator's descriptions of the their respective vessels.

455. **e.** The captain knew it was important that the crew understood the boy was no more privileged than anyone else aboard the *Alba*. Evidence for this choice is found in the narrator's statement in lines 10–11: *because I was his nephew, I would probably have to work twice as hard as the others to prove my worth.* All the other choices do not make sense or are not backed by specific examples from the text.

456. **a.** As used inPassage 1, line 26, the verb *repair* most closely means take themselves, or more simply, go. Today, repair is most commonly used as a verb that means to fix something (choice **b**). However, in the context of the sentence, this makes no sense. The easiest way to answer this question is to replace *repair* in the sentence with each the answer choices, and see which one fits best in context. By doing this you should narrow down your choice to just one: choice **a**.

457. **c.** The narrator's familiarity with yachts and the harbor at Plymouth (lines 1–12) in Passage 1 seems to indicate that he is an experienced yachtsman. He reveals his passion for yachting in lines 17–18, when he declares, *Of all amusements, give me yachting.* All the other answer choices either do not make sense or are not supported by specific examples from the text.

458. **e.** Nigel probably had rotten or missing teeth. The narrator of Passage 2 chose to describe Nigel's smile as *a graveyard of yellow sincerity*, describing his yellow teeth as tombstones in a graveyard. When a writer uses a descriptive word or phrase in place of another to suggest a similarity between the two, this figure of speech is called a *metaphor* (choice **e**). If the boy had instead said, Nigel's smile was "*like* a graveyard of yellow sincerity," it would have been a *simile*, choice **b**.

459. **a.** Both passages are basically concerned with a similar situation—life aboard a cutter. The author of Passage 1 sets a pleasurable tone in the first paragraph by describing the idyllic scene at Plymouth and the anchored yacht. He later describes the yacht as *elegant, tasteful,* and *luxurious* (line 18), and the smell of the food *delightful* (lines 23–24). In stark contrast, the boy narrator in Passage 2 begins the passage by describing the menacing façade of his uncle and the immediate reality check the boy receives when he steps aboard (lines 6–9). His description of the heat and smell below deck (lines 20–22), and the horrible food (lines 33–35), effectively sets the dark and oppressive tone of the passage. Together, these two very different descriptions prove that the *reality of two seemingly similar situations can often be extremely different*, choice **a**.

460. **b.** The word *berth*, when used as a noun, often refers to the sleeping quarters aboard a boat or a train. In lines 39–43 the boy describes his berth as the place where he could *stow [his] clothes,* and at night *string up [his] hammock.*

Social Studies

Questions 461–464 are based on the following passage.

The following passage examines the possibility that early humans used toothpicks.

(1)   Could good dental hygiene be man's earliest custom? The findings of paleontologist Leslea Hlusko suggest that 1.8 million years ago early hominids used grass stalks to clean their teeth. Many ancient hominid teeth unearthed in archaeological digs have curved grooves near the
(5)   gumline. Hlusko posited that these grooves were evidence of teeth cleaning by early man. However, critics pointed out that even though the use of toothpicks is still a common practice among modern man similar grooves are not found on modern teeth.

      Hlusko, convinced that she was on the right track, experimented
(10)   with grass stalks to see if they might have been the cause of the grooves. Unlike the wood used for modern toothpicks, grass contains hard silica particles that are more abrasive than the soft fibers found in wood. A stalk of grass is also about the same width as the marks found on the ancient teeth. To prove her theory Dr. Hlusko took a
(15)   baboon tooth and patiently rubbed a grass stalk against it for eight hours. As she suspected, the result was grooves similar to those found on the ancient hominid teeth. She repeated the experiment with a human tooth and found the same result.

It seems that our early human ancestors may have used grass, which
(20) was easily found and ready to use, to floss between their teeth. As
Hlusko suggests in the journal *Current Anthropology,* "Toothpicking
with grass stalks probably represents the most persistent habit documented in human evolution."

461. In line 5 the word *posited* most nearly means
   a. insisted.
   b. demanded.
   c. questioned.
   d. suggested.
   e. argued.

462. Each of the following reasons is provided as evidence that early
   man used grass stalks as toothpicks EXCEPT the
   a. width of the grooves on ancient teeth.
   b. location of the grooves on ancient teeth.
   c. ready availability of grass.
   d. ongoing use of grass toothpicks.
   e. abrasive quality of grass.

463. Dr. Hlusko's approach to determining the source of the grooves on
   ancient teeth can best be described as
   a. zealous.
   b. persistent.
   c. sullen.
   d. serendipitous.
   e. cautious.

464. The passage suggests the theory that early man used grass stalks as
   toothpicks is
   a. a possibility.
   b. very probable.
   c. absolutely certain.
   d. fanciful.
   e. uncorroborated.

Questions 465–469 are based on the following passage.

The following passage analyzes data from the U.S. Census Bureau to draw conclusions about the economic well being of Americans in the years 1993 and 1994.

(1)   From year to year, the economic well being of many Americans changes considerably, even though the median income of the population as a whole does not vary much in real terms from one year to the next. One measure of economic well being is the income-to-

(5)   poverty ratio. This ratio measures a family's income compared to the poverty threshold (the income below which a family is considered to be in poverty) for that family. For example, the poverty threshold for a three-person family in 1994 was $11,817. A three-person family with an income of $20,000 would have an income-to-

(10)   poverty ratio of 1.69 ($\frac{\$20,000}{\$11,817}$).

        Between 1993 and 1994 roughly three-quarters of the population saw their economic well being fluctuate by 5% or more. Conversely, from year to year less than a quarter of Americans had stable incomes. In the 1990s fewer people saw their income grow than in the 1980s,

(15)   and more people saw their incomes decline. Although the state of the economy is a notable factor in determining if incomes rise or fall, changes in personal circumstances are just as important. People had a good chance of seeing their income rise if they began to work full-time, the number of workers or adults in their house increased, they

(20)   married, or the number of children in the household decreased. Conversely, people could expect a decrease in their income if they ceased to be married or to work full-time.

        Another factor that affected the direction of change in family income was its place on the economic ladder. The closer a family was

(25)   to poverty the more likely they were to see their income rise. Whereas, 45% of families at the top of the economic ladder, those with income-to-poverty ratios of more than 4.0, experienced income decreases in 1994. While age, gender, and race play a significant role in determining one's place on the economic ladder, these factors are

(30)   not good predictors of a rise or fall in income. The only population for which one of these factors was significant was the elderly, whose incomes tended to be fairly stable.

465. According to the passage, in general, income across the United States tends to
   a. fluctuate wildly.
   b. change incrementally.
   c. increase slightly.
   d. decrease steadily.
   e. stay about the same.

466. The first paragraph of the passage serves all the following purposes EXCEPT to
   a. define the term *poverty threshold*.
   b. explain income-to-poverty ratio.
   c. provide an example of an income-to-poverty ratio.
   d. state the author's thesis.
   e. establish the subject of the passage.

467. According to the passage, people's income in the 1990s was
   a. likely to rise.
   b. likely to fall.
   c. greater than in the 1980s.
   d. less than in the 1980s.
   e. less likely to grow than in the 1980s.

468. In the context of this passage, the phrase *the economic ladder* (line 26) most nearly means
   a. the range of occupations.
   b. the pecking order.
   c. the capitalist social structure.
   d. the caste system.
   e. the range of incomes.

469. The tone of this passage can best be described as
   a. dry and neutral.
   b. statistical.
   c. unintentionally witty.
   d. theoretical.
   e. inflammatory.

Questions 470–476 are based on the following passage.

This passage, from research conducted for the Library of Congress Folklife Center, discusses the various folk beliefs of Florida fishermen.

(1)　Beliefs are easily the most enduring and distinctive aspects of maritime culture. Traditional beliefs, commonly called superstitions, are convictions that are usually related to causes and effects, and are often manifest in certain practices. Common examples include beliefs about
(5)　good and bad luck, signs for predicting the weather, interpretations of supernatural happenings, and remedies for sickness and injury.

　　Because maritime occupations often place workers in a highly unpredictable and hazardous environment, it is not surprising that fishermen hold many beliefs about fortune and misfortune. A primary function of
(10)　such beliefs is to explain the unexplainable. Watermen can cite many actions that invite bad luck. These actions include uttering certain words while aboard a boat, taking certain objects aboard a boat, going out in a boat on a certain day, or painting boats certain colors. Among Florida fishermen, saying "alligator," bringing aboard shells or black suitcases,
(15)　and whistling are all considered bad luck while on a boat.

　　Beliefs about actions that invite good luck appear to be fewer in number than those about bad luck. Beliefs about good luck include breaking a bottle of champagne or other liquid over the bow of a vessel when it is launched, participating in a blessing-of-the-fleet cere-
(20)　mony, placing a coin under the mast, carrying a lucky object when aboard, and stepping on or off the boat with the same foot. There are many beliefs about predicting the weather and the movement of fish. These beliefs are often linked to the detection of minute changes in the environment and reflect fishermen's intimate contact with the nat-
(25)　ural environment.

　　A Florida shrimp fisherman told a researcher that when shrimps' legs are blood red you can expect a strong northeaster or strong southeaster. The direction of the wind is used to predict the best location for catching shrimp. Other signs for weather prediction include rings
(30)　around the moon, the color of the sky at sunrise and sunset, and the color and texture of the sea. Sometimes beliefs are expressed in concise rhymes. An oysterman from Apalachicola, Florida, uses the rhyme, "East is the least, and west in the best" to recall that winds from the west generally produce conditions that are conducive to good catches.
(35)　Beliefs related to the supernatural—the existence of ghosts, phantom ships, burning ships, or sea monsters—are also found in maritime communities. Many fishermen are reluctant to discuss the supernatural, so these beliefs are less conspicuous than those about luck and the

(40) weather. However, one net maker told a researcher about his encounter with a ghost ship. He saw a schooner, a ship that was prevalent in the nineteenth century, come in across the Gulf and pass through water that was far too shallow for a ship of its size. The ship then suddenly disappeared from sight.

Commercial fishing is considered to be the most hazardous of all (45) industrial occupations in the United States. Statistics show that fishermen are seven times more likely to die than workers in the next most dangerous occupation. Adhering to a system of beliefs most likely helps bring sense and order to a world in which natural disasters and misfortune are a part of daily life. Many fishermen also make a pre-(50) carious living at best. Maritime beliefs contain the collective wisdom of generations and following these traditions may help fishermen catch more fish without taking unnecessary risks.

470. In line 4, the phrase *manifest in certain practices* most nearly means
  a. obviously rehearsed.
  b. recorded in some religions.
  c. destined in certain circumstances.
  d. evident in particular activities.
  e. decreed in unwavering terms.

471. According to the passage, fishermen are superstitious because
  a. they learn it from previous generations.
  b. they believe in the supernatural.
  c. fishing is a dangerous and unpredictable occupation.
  d. they are afraid of stormy weather.
  e. fishing is a terrible way to make a living.

472. The author's attitude toward fishermen's beliefs about predicting the weather can best be characterized as
  a. unqualified respect.
  b. veiled disbelief.
  c. tempered belief.
  d. absolute fascination.
  e. minimal enthusiasm.

473. According to information in the passage, fishermen's beliefs about the supernatural do not conform to the author's definition of traditional beliefs (lines 2–4) in that
  a. fishermen do not like to talk about them.
  b. they are not related to cause and effect.
  c. they are not conspicuous.
  d. they are not manifest.
  e. they are less rooted in the natural world.

474. The purpose of the statistic in lines 45–47 is to
  a. qualify the statement that fishing is hazardous.
  b. prove that fishing is an undesirable occupation.
  c. illustrate the relative ease of other professions.
  d. quantify the hazardous nature of commercial fishing.
  e. demonstrate that fishermen need a system of beliefs.

475. In lines 49–50, *precarious* most nearly means
  a. dangerous.
  b. steady.
  c. reduced.
  d. meager.
  e. uncertain.

476. The primary purpose of the passage is to
  a. catalog the beliefs of Florida fishermen.
  b. demonstrate that traditional beliefs are effective.
  c. describe some traditional beliefs found among Florida fishermen.
  d. prove that superstitions are a valid guide to behavior.
  e. amuse readers with the peculiar beliefs of Florida fishermen.

Questions 477–483 are based on the following passage.

This passage explores the theory that the first three years of life are critical in the development of a child's character and suggests a parenting model that strengthens moral behavior.

(1)   Does a baby have a moral conscience? While a baby is not faced with many serious ethical dilemmas, his or her moral character is formed from the earliest stages of infancy. Recent research has shown that the type of parenting an infant receives has a dramatic impact on the
(5)   child's moral development and, consequently, success later in life. The renowned childcare expert T. Berry Brazelton claims that he can

observe a child of eight months and tell if that child will succeed or fail in life. This may be a harsh sentence for an eight-month-old baby, but it underscores the importance of educating parents in good child-rear-
(10) ing techniques and of intervening early in cases of child endanger-ment. But what are good parenting techniques?

The cornerstone of good parenting is love, and the building blocks are trust, acceptance, and discipline. The concept of "attachment par-enting" has come to dominate early childhood research. It is the rela-
(15) tively simple idea that an infant who is firmly attached to his or her "primary caregiver"—often, but not always, the mother—develops into a secure and confident child. Caregivers who respond promptly and affectionately to their infants' needs—to eat, to play, to be held, to sleep, and to be left alone—form secure attachments with their children. A
(20) study conducted with rhesus monkeys showed that infant monkeys pre-ferred mothers who gave comfort and contact but no food to mothers who gave food but no comfort and contact. This study indicates that among primates love and nurturing are even more important than food.

Fortunately, loving their infants comes naturally to most parents and
(25) the first requisite for good parenting is one that is easily met. The sec-ond component—setting limits and teaching self-discipline—can be more complicated. Many parents struggle to find a balance between responding promptly to their babies' needs and "spoiling" their child. Norton Garfinkle, chair of the Executive Committee of the Lamaze
(30) Institute for Family Education, has identified four parenting styles: warm and restrictive, warm and permissive, cold and restrictive, and cold and permissive. A warm parent is one who exhibits love and affection; a cold parent withholds love; a restrictive parent sets limits on her child's behav-ior and a permissive parent does not restrict her child. Garfinkle finds
(35) that the children of warm-restrictive parents exhibit self-confidence and self-control; the children of warm-permissive parents are self-assured but have difficulty following rules; children of cold-restrictive parents tend to be angry and sullenly compliant, and the most troubled children are those of cold-permissive parents. These children are hostile and defiant.
(40) The warm-restrictive style of parenting helps develop the two key dimensions of moral character: empathy and self-discipline. A warm attachment with his or her parent helps the child develop empathetic feelings about other human beings, while parental limit-setting teaches the child self-discipline and the ability to defer gratification.
(45) The ability to defer gratification is an essential skill for negotiating the adult world. A study conducted by Daniel Goleman, author of *Emo-tional Intelligence*, tested a group of four-year-olds' ability to defer grat-ification. Each child in the study was offered a marshmallow. The

(50) child could choose to eat the marshmallow right away or wait fifteen minutes to eat the marshmallow and receive another marshmallow as a reward for waiting. Researchers followed the children and found that by high school those children who ate their marshmallow right away were more likely to be lonely, more prone to stress, and more easily frustrated. Conversely, the children who demonstrated self-control

(55) were outgoing, confident, and dependable.

This research seems to answer the old adage, "you can't spoil a baby." It seems that a baby who is fed at the first sign of hunger and picked up on demand can perhaps be "spoiled." Most parents, how-ever, tend to balance their baby's needs with their own. Many parents

(60) will teach their baby to sleep through the night by not picking up the baby when she awakes in the middle of the night. Although it can be heart wrenching for these parents to ignore their baby's cries, they are teaching their baby to fall asleep on her own and getting the benefit of a full night's sleep.

(65) While many parents will come to good parenting techniques instinctually and through various community supports, others parents are not equipped for the trials of raising a baby. Are these babies doomed to lives of frustration, poor impulse-control, and anti-social behavior? Certainly not. Remedial actions—such as providing enrich-

(70) ment programs at daycare centers and educating parents—can be taken to reverse the effects of bad parenting. However, the research indicates that the sooner these remedies are put into action the better.

477. The primary purpose of the passage is to
    a. advocate for the ability to defer gratification.
    b. educate readers about moral development in infants.
    c. chastise parents for spoiling their children.
    d. inform readers of remedies for bad parenting.
    e. demonstrate the importance of love in child rearing.

478. In line 8, the word *sentence* most nearly means
    a. statement.
    b. pronouncement.
    c. declaration.
    d. judgment.
    e. punishment.

479. The author presents the study about rhesus monkeys
(lines 19–22) to
    a. prove that humans and monkeys have a lot in common.
    b. suggest that food is used as a substitute for love.
    c. support her assertion that love is the most important aspect of
good parenting.
    d. disprove the idea that you can't spoil a baby.
    e. broaden the scope of her argument to include all primates.

480. According to the third paragraph of the passage, a cold-restrictive
parent can best be characterized as
    a. an aloof disciplinarian.
    b. an angry autocrat.
    c. a frustrated teacher.
    d. a sullen despot.
    e. an unhappy dictator.

481. Based on the information in paragraph four, one can infer that
children who are unable to defer gratification are most unlikely to
succeed because
    a. they are unpopular.
    b. they lack empathy.
    c. their parents neglected them.
    d. they are unable to follow directions.
    e. they lack self-discipline.

482. Which of the following techniques is used in lines 59–64?
    a. explanation of terms
    b. comparison of different arguments
    c. contrast of opposing views
    d. generalized statement
    e. illustration by example

483. The author of this passage would be most likely to agree with
which statement?
    a. Babies of cold-permissive parents are doomed to lives of failure.
    b. Good parenting is the product of education.
    c. Instincts are a good guide for most parents.
    d. Conventional wisdom is usually wrong.
    e. Parents should strive to raise self-sufficient babies.

2-23-14

Questions 484–492 are based on the following two passages.

Passage 1 describes the potlatch ceremony celebrated by native peoples of the Pacific Northwest. Passage 2 describes the kula ring, a ceremonial trading circle practiced among Trobriand Islanders in Papua New Guinea.

PASSAGE 1

(1)   Among traditional societies of the Pacific Northwest—including the Haidas, Kwakiuls, Makahs, Nootkas, Tlingits, and Tsimshians—the gift-giving ceremony called potlatch was a central feature of social life. The word *potlatch*, meaning "to give," comes from a Chinook trading

(5)   language that was used all along the Pacific Coast. Each nation, or tribe, had its own particular word for the ceremony and each had different potlatch traditions. However, the function and basic features of the ceremony were universal among the tribes.

          Each nation held potlatches to celebrate important life passages,

(10)   such as birth, coming of age, marriage, and death. Potlatches were also held to honor ancestors and to mark the passing of leadership. A potlatch, which could last four or more days, was usually held in the winter when the tribes were not engaged in gathering and storing food. Each potlatch included the formal display of the host family's crest and

(15)   masks. The hosts performed ritual dances and provided feasts for their guests. However, the most important ritual was the lavish distribution of gifts to the guests. Some hosts might give away most or all of their accumulated wealth in one potlatch. The more a host gave away, the more status was accorded him. In turn, the guests, who had to accept

(20)   the proffered gifts, were then expected to host their own potlatches and give away gifts of equal value.

          Prior to contact with Europeans, gifts might include food, slaves, copper plates, and goat's hair blankets. After contact, the potlatch was fundamentally transformed by the influx of manufactured goods. As

(25)   tribes garnered wealth in the fur trade, gifts came to include guns, woolen blankets, and other Western goods. Although potlatches had always been a means for individuals to win prestige, potlatches involving manufactured goods became a way for nobles to validate tenuous claims to leadership, sometimes through the destruction of property. It

(30)   was this willful destruction of property that led Canadian authorities, and later the U.S. government, to ban potlatches in the late 1880s.

          Despite the ban, the potlatch remained an important part of native Pacific Northwest culture. Giving wealth—not accumulating wealth, as is prized in Western culture—was a means of cementing leadership,

(35) affirming status, establishing and maintaining alliances, as well as ensuring the even distribution of food and goods. Agnes Alfred, an Indian from Albert Bay, explained the potlatch this way, "When one's heart is glad, he gives away gifts. . . . The potlatch was given to us to be our way of expressing joy."

PASSAGE 2

(1) The inhabitants of the Trobriand Islands, an archipelago off the coast of Papua New Guinea in the South Pacific, are united by a ceremonial trading system called the kula ring. Kula traders sail to neighboring islands in large ocean-going canoes to offer either shell necklaces *(5)* or shell armbands. The necklaces, made of red shells called *bagi*, travel around the trading ring clockwise, and the armbands, made of white shells called *mwali*, travel counterclockwise.

Each man in the kula ring has two kula trading partners—one partner to whom he gives a necklace for an armband of equal value, *(10)* although the exchanges are made on separate occasions, and one partner with whom he makes the reverse exchange. Each partner has one other partner with whom he trades, thus linking all the men around the kula ring. For example, if A trades with B and C, B trades with A and D, and C trades with A and E, and so on. A man may have only *(15)* met his own specific kula partners, but he will know by reputation all the men in his kula ring. It can take anywhere from two to ten years for a particular object to complete a journey around the ring. The more times an object has made the trip around the ring the more value it accrues. Particularly beautiful necklaces and armbands are also *(20)* prized. Some famous kula objects are known by special names and through elaborate stories. Objects also gain fame through ownership by powerful men, and, likewise, men can gain status by possessing particularly prized kula objects.

The exchange of these ceremonial items, which often accompanies *(25)* trade in more mundane wares, is enacted with a host of ritual activities. The visitors, who travel to receive kula from their hosts, are seen as aggressors. They are met with ritual hostility and must charm their hosts in order to receive the necklaces or armbands. The visitors take care to make themselves beautiful, because beauty conveys strength *(30)* and protects them from danger. The hosts, who are the "victims" of their visitors' charm and beauty, give the prized objects because they know that the next time it will be their turn to be the aggressor. Each man hopes that his charm and beauty will compel his trading partner to give him the most valuable kula object.

(35)   The objects cannot be bought or sold. They have no value other than their ceremonial importance, and the voyages that the traders make to neighboring islands are hazardous, time-consuming, and expensive. Yet, a man's standing in the kula ring is his primary concern. This ceremonial exchange has numerous tangible benefits. It estab-
(40) lishes friendly relations through a far-flung chain of islands; it provides a means for the utilitarian exchange of necessary goods; and it reinforces the power of those individuals who win and maintain the most valuable kula items. Although the kula ring might mystify Western traders, this system, which has been in operation for hundreds of
(45) years, is a highly effective means of unifying these distant islanders and creating a common bond among peoples who might otherwise view one another as hostile outsiders.

484. According to Passage 1, *potlatch* is best defined as a
    a. ceremony with rigid protocol to which all Pacific Northwest tribes adhere.
    b. generic term for a gift-giving ceremony celebrated in the Pacific Northwest.
    c. socialist ritual of the Pacific Northwest.
    d. lavish feast celebrated in the Pacific Northwest.
    e. wasteful ritual that was banned in the 1880s.

485. According to Passage 1, the gift-giving central to the potlatch can best be characterized as
    a. reciprocal.
    b. wasteful.
    c. selfless.
    d. spendthrift.
    e. commercialized.

486. In Passage 1, the author's attitude toward the potlatch can best be described as
    a. condescending.
    b. antagonistic.
    c. wistful.
    d. respectful.
    e. romantic.

487. According to Passage 2, the men in a kula ring are
   a. linked by mutual admiration.
   b. hostile aggressors.
   c. greedy.
   d. motivated by vanity.
   e. known to one another by reputation.

488. In Passage 2, line 30, the word *victims* is in quotation marks because the
   a. word might be unfamiliar to some readers.
   b. author is implying that the hosts are self-pitying.
   c. author is reinforcing the idea that the hosts are playing a pre-scribed role.
   d. author wants to stress the brutal nature of the exchange.
   e. author is taking care not to be condescending to the Trobriand culture.

489. According to Passage 2, necklaces and armbands gain value through all the following means EXCEPT being
   a. in circulation for a long time.
   b. especially attractive.
   c. owned by a powerful man.
   d. made of special shells.
   e. known by a special name.

490. Gift-giving in the potlatch ceremony and the ritual exchange of the kula ring are both
   a. a ritualized means of maintaining community ties.
   b. dangerous and expensive endeavors.
   c. a means of ascending to a position of leadership.
   d. falling prey to Western culture.
   e. peculiar rituals of a bygone era.

491. Based on information presented in the two passages, both authors would be most likely to agree with which statement?
   a. Traditional societies are more generous than Western societies.
   b. The value of some endeavors cannot be measured in monetary terms.
   c. It is better to give than to receive.
   d. Westerners are only interested in money.
   e. Traditional societies could benefit from better business sense.

492. Which of the following titles would be most appropriate for both Passage 1 or Passage 2?
   a. A Gift-giving Ceremony
   b. Ritual Exchange in Traditional Societies
   c. Ceremonial Giving and Receiving in a Traditional Society
   d. The Kindness of Strangers
   e. Giving and Receiving in a Faraway Land

2-24-14

Questions 493–501 are based on the following passage.

The author of this passage, a professor of English literature at a major university, argues that affirmative action is a necessary part of the college admissions process.

(1)   When I began teaching at Big State U in the late 1960s, the students in my American literature survey were almost uniformly of European heritage, and most were from middle-class Protestant families. Attending college for these students was a lesson in homogeneity.

(5)   Although a number of students were involved in the Civil Rights Movement and some even worked "down South" on voter registration, most students considered segregation to be a Southern problem and many did not see the discrimination that was rampant on their own campus.

(10)   Since the 1960s there has been a sea change in university admissions. Key Supreme Court decisions and federal laws made equal opportunity the law of the land, and many institutions of higher learning adopted policies of affirmative action. The term *affirmative action* was first used in the 1960s to describe the active recruitment and pro-

(15)   motion of minority candidates in both the workplace and in colleges and universities. President Lyndon Johnson, speaking at Howard University in 1965, aptly explained the reasoning behind affirmative action. As he said, "You do not take a man who, for years, has been hobbled by chains and liberate him, bring him to the starting line in

(20)   a race and then say, 'You are free compete with all the others,' and still believe that you have been completely fair." Affirmative action programs in college admissions have been guided by the principle that it is not enough to simply remove barriers to social mobility but it is also necessary to encourage it for minority groups.

(25)   In recent years, affirmative action programs have come under public scrutiny, and some schools have been faced with charges of reverse discrimination. Preferential treatment of minority applicants is seen as discrimination against qualified applicants from the majority

group. Despite widespread support for the elimination of prejudice,
(30) most whites do not favor the preferential treatment of minority applicants, and affirmative action in college admissions has been abolished
in several states. In my view, this trend is very dangerous not only for
minority students but for all students. Thanks to a diversified student
body, my classes today are much richer than when I began teaching in
(35) the 1960s. For example, when I teach *A Light in August* by William
Faulkner, as I do every fall, today there is likely to be a student in the
class who has firsthand knowledge of the prejudice that is a central
theme of the novel. This student's contribution to the class discussion
of the novel is an invaluable part of all my students' education and a
(40) boon to my experience as a teacher.

Some may argue that affirmative action had its place in the years
following the Civil Rights Movement, but that it is no longer necessary. To assume that all students are now on a level playing field is
naïve. Take for example the extra-curricular activities, AP classes, and
(45) internships that help certain applicants impress the admissions board:
These are not available or economically feasible for many minority
candidates. This is just one example of why affirmative action still has
an important place on American campuses. When all things are equal,
choosing the minority candidate not only gives minorities fair access
(50) to institutions of higher learning, but it ensures diversity on our campuses. Exposing all students to a broad spectrum of American society
is a lesson that may be the one that best prepares them to participate
in American society and succeed in the future.

493. In line 4, the phrase *a lesson in homogeneity* can be most accurately
described as
   a. a slight against civil rights workers.
   b. an ironic observation about the uniform character of the student body.
   c. a comment on the poor quality of the education at Big State U.
   d. a sarcastic comment about the authors' former students.
   e. the author's' rueful view of his poor teaching skills.

494. In line 10, the expression *sea change* means
   a. increase.
   b. storm.
   c. decrease.
   d. wave.
   e. transformation.

495. The author uses the quote from President Lyndon Johnson in (lines 18–21) to
   a. provide an example of discrimination in the past.
   b. show how Howard University benefited from affirmative action policies.
   c. make the passage more interesting.
   d. explain the rationale for affirmative action.
   e. prove that affirmative action has been effective at promoting diversity.

496. According to the passage, the greatest danger of abolishing affirmative action in college admissions is
   a. allowing reverse discrimination to take hold of college admissions.
   b. creating a "slippery slope" of discrimination and prejudice.
   c. losing the benefits of a diverse campus.
   d. returning to the segregation of the past.
   e. complicating the job of the college admissions board.

497. From the information provided in the passage, one can conclude that the author
   a. has personally benefited from the effects of affirmative action.
   b. considers affirmative action a necessary evil.
   c. favors accepting poorly qualified candidates for the sake of diversity.
   d. despises the opponents of affirmative action.
   e. thinks that affirmative action will eventually be unnecessary.

498. The word *feasible* in line 46 most nearly means
   a. advantageous.
   b. possible.
   c. attractive.
   d. probable.
   e. suitable.

499. The tone of this passage can best be described as
   a. impassioned.
   b. impartial.
   c. reasonable.
   d. sarcastic.
   e. dispassionate.

500. The author gives all the following reasons for continuing affirmative action in college admissions EXCEPT that it
   a. fosters diversity.
   b. provides fair access to higher education.
   c. is necessary to promote social mobility.
   d. exposes students to a broad spectrum of society.
   e. prepares students for the future.

501. The argument for affirmative action in the workplace that most closely mirrors the author's reasoning about affirmative action in college admissions is
   a. it is the law of the land.
   b. diversity in the workplace better prepares a company to compete in the marketplace.
   c. a diverse workforce is more efficient.
   d. a less-qualified minority candidate is still a great asset to a company.
   e. it is the right thing to do.

Answers

461. **d.** To *posit* means to suggest. In this context, Hlusko suggests that grass stalks may have caused the grooves on early hominid teeth.

462. **d.** The passage states that modern toothpicks are made of wood (line 11).

463. **b.** Dr. Hlusko is described a being *convinced she was on the right track* and *patiently* rubbing a baboon tooth with a grass stalk for eight hours. Both point to a persistent approach.

464. **b.** In lines 19–20, the author states, *It seems that our early human ancestors may have used grass, which was easily found and ready to use, to floss between their teeth.* The use of *may* indicates that the author is not *absolutely certain*, but as the author does not suggest anything to contradict Dr. Hlusko's findings we can conclude that the author finds her theory very probable.

465. **e.** The passage clearly states that *the median income of the population as a whole does not vary much in real terms from one year to the next.* From this statement one can infer that, in general, income across the United States stays about the same.

466. **d.** A thesis is an assertion, or theory, that the author intends to prove. The author of this passage is not making an assertion, rather he or she is neutrally explaining information gathered in the U.S. Census.

467. e. The passage clearly states that in the 1990s *fewer people saw their income grow than in the 1980s.* Choices **a** and **b** are incorrect because they do not include a comparison to the 1980s. Choices **c** and **d** are incorrect because the passage does not discuss *amount* of income, only change in income.

468. e. The passage defines *top of the economic ladder* as families with high income-to-poverty ratios. From this, one can conclude that the economic ladder is the range of incomes from poverty to wealth.

469. a. The tone is dry, in that the language is spare. The author does not use many adjectives, or any metaphors or other rhetorical flourishes. The author is neutral. Nowhere in the passage does he or she assert a point of view. Although the author uses statistics, the tone is not most accurately described as statistical.

470. d. *Manifest* means obvious or evident. *Certain practices* could have several meanings and it is necessary to look to the examples provided in the next sentence to clarify the meaning of the phrase. The examples of beliefs mostly relate to particular activities such as predicting the weather or curing sickness.

471. c. Lines 7–9 clearly state that it is not surprising that fishermen hold many beliefs about fortune and misfortune because fishermen work in a *highly unpredictable and hazardous environment.*

472. a. In lines 21–25, the author states that these beliefs are *linked to the detection of minute changes in the environment* and *reflect fishermen's intimate contact with the natural environment.* This sentence indicates an attitude of respect. This respect is unqualified in that the author does not detract from the statement in any way.

473. b. The author defines *traditional beliefs* as *convictions that are usually linked to causes and effects.* In the paragraph that discusses supernatural (lines 35–43), the author states that some fishermen believe in the *existence* of the certain supernatural phenomena. There is no information about the cause or effect of the supernatural.

474. d. The statistic provides numerical evidence (quantifies) of the degree to which commercial fishing is hazardous compared to the next most dangerous occupation.

475. e. *Precarious* means dependent on uncertain circumstances or chance; it can also mean characterized by a lack of security. Uncertain and dangerous (choice **a**) are both synonyms of *precarious*, however, in the context of the sentence, uncertain makes the most sense.

476. **c.** The passage is primarily concerned with describing beliefs found among Florida fishermen. The passage does not, however, catalog (give a complete account of) their beliefs, in that it only gives some examples. Although the author does close the passage with a suggestion traditional beliefs may have some real world benefits, the majority of the passage is not occupied with this idea.

477. **b.** The primary purpose of the passage is to educate readers about the importance of good parenting in developing moral character in children. Choices **a**, **d**, and **e** are too narrow. Choice **c** is not supported by the passage.

478. **d.** The author is using *sentence* in the sense of a conclusion reached by a judge in a criminal trial. She is asserting that to conclude that an eight-month-old baby is already destined for success or failure is a harsh judgment on such a small child. Note that choice **e**, is incorrect because punishment is the result of a sentence, and does not make sense in this context.

479. **c.** The author opens the paragraph with the assertion that love is the *cornerstone* (foundation) of good parenting. The monkey study, which indicates that the need for love supercedes the need for food, is used to support that assertion.

480. **a.** The passage clearly defines *cold* parents as withholding love (lines 32–33). *Aloof* means reserved or removed in feeling. *Restrictive* parenting is defined in the passage as setting limits (lines 33–34). A *disciplinarian* is one who enforces order.

481. **e.** Lines 44 and 54–55 link the ability to defer gratification with self-discipline and self-control. Hence, children who are unable to defer gratification are unlikely to succeed because they lack self-discipline.

482. **e.** The subject of this paragraph is parents balancing their needs with those of their child. Teaching a child to sleep through the night is an example of parents balancing their needs (for a full night's sleep) with the needs of their baby (to be picked up in the middle of the night).

483. **c.** The passage clearly states that *many parents will come to good parenting techniques instinctually* (lines 65–66), which indicates that instincts are a good guide for parents. Also, line 24 states that loving an infant comes naturally to most parents—something that comes *naturally* is instinctual. None of the other choices is supported by the passage.

484. **b.** The passage clearly states that *potlatch* is a gift-giving ceremony. The author explains that *potlatch* is a generic word for the cere-

mony that comes from a shared trading language, while each
nation has its own specific word for *potlatch*.

485. **a.** The passage states that guests were expected to give a potlatch
with gifts of equal value to what they received. This arrangement can best be described as reciprocal. The other choices are
not supported by the passage.

486. **d.** The author describes the ceremony in mostly neutral terms but
in the last paragraph emphasizes the positive aspects of the tradition, which indicates a degree of respect.

487. **e.** The passage explicitly states in lines 15–16 that a man *will know
by reputation all the men in his kula ring*. None of the other
choices is explicitly stated in the passage.

488. **c.** The passage states in lines 26–27 that the visitors are *seen as
aggressors* and are met with *ritual hostility*. This indicates that the
visitors and hosts are playing the roles of aggressor and victims.
The author uses quotes to indicate that the hosts are not really
victims, but might call themselves the *victims* in the exchange.

489. **d.** Lines 17–24 state the ways in which a kula object gains value;
special shells are not mentioned.

490. **a.** The final paragraph of each passage explicitly states the ways in
which these ceremonies, or rituals, maintain community ties.
None of the other choices is true for both passages.

491. **b.** Both authors specifically discuss the non-monetary value of
each ceremony. In Passage 1, lines 33–36 the author states, *Giving wealth—not accumulating wealth, as is prized in Western culture—was a means of cementing leadership, affirming status, . . .* In
Passage 2, lines 35–39 the author states, *The objects . . . have no
value*, and yet, *this ceremonial exchange has numerous tangible benefits*. None of the other choices is supported by the texts.

492. **c.** Both potlatches and the kula ring involve giving and receiving,
and both of the societies that participate in these rituals can be
described as traditional. The tone of the title in choice **e** is
more whimsical than the serious tone of each passage. Choice **b**
is incorrect because neither article draws conclusions about traditional societies in general.

493. **b.** The sentence preceding this phrase discusses the homogenous,
or uniform, makeup of the student body in the 1960s. The
author is using the word *lesson* ironically in that a lack of diversity is not something on which many educators would pride
themselves.

494. **e.** A *sea change* is a transformation. This can be inferred from the
next sentence, which states that colleges adopted policies of

affirmative action. Affirmative action is a transformation in college admissions.

495. **d.** The author clearly states in lines 17–18 that President Johnson *aptly explained the reasoning behind affirmative action.*

496. **c.** After stating that he considers the trend of abolishing affirmative action to be very dangerous, the author explains how a diverse student body makes his classes *much richer.*

497. **a.** According to the author, one of the main benefits of affirmative action is diversity in the classroom and he states that this diversity has been *a boon to my experience as a teacher* (line 40). So, affirmative action has personally benefited the author. None of the other choices is supported by the passage.

498. **b.** *Feasible* can mean capable of being done (possible) or capable of being used (suitable). In this context, the author is suggesting that, for many minorities, extracurricular activities and the like are not economically possible, that is they are unaffordable.

499. **c.** The author expresses his opinion about affirmative action in a moderate, or reasonable, tone. He is neither dispassionate nor passionate, in that he expresses some emotion but not much. He is not impartial, as he is expressing an opinion.

500. **e.** It is diversity, the result of affirmative action, not affirmative action itself, that prepares students for the future (lines 51–53).

501. **b.** The author's main argument for affirmative action is that the student body benefits from diversity. His final point is that students who have been exposed to *a broad spectrum of American society* (line 51) are better prepared for their futures. The idea that diversity benefits a company and makes it better prepared to compete in marketplace most closely mirrors this reasoning.

Source Materials

U.S. History and Politics

Pages 27–28: *Abraham Lincoln Papers*, Library of Congress Manuscript Division.

Pages 32–33: *The African American Odyssey: A Quest for Full Citizenship*, Library of Congress, www.memory.loc.gov. (Adapted.)

Pages 35–36: *The Chinese in California, 1850–1925*, Library of Congress, University of Berkeley, California, and the California Historical Society, www.memory.loc.gov.

Pages 38–39: National Park Service, Department of the Interior, Lowell National Historical Park, text by Thomas Dublin.

Pages 45–46: *Rivers, Edens, Empires: Lewis & Clark and the Revealing of America*, Library of Congress, www.loc.gov/exhibits/lewisandclark/lewisandclark.html.

Pages 49–51: Library of Congress, Rare Book and Special Collections Division, National American Woman Suffrage Association Collection.

Arts and Humanities

Pages 60–61: *Nicomachean Ethics*. Aristotle. Translated by Martin Ostwald. NY: Macmillan, 1962.

Health and Medicine

Pages 87–88: National Institute of Neurological Disorders and Stroke, National Institutes of Health, www.ninds.nih.gov.

Pages 90–91: National Library of Medicine, www.nlm.nih.gov.

Pages 99–101: National Library of Medicine, http://profiles.nlm.nih.gov.

Literature and Literary Criticism

Pages 121–122: *Angela's Ashes.* McCourt, Frank. NY: Scribner, 1996.

Page 123: *The Bluest Eye.* Morrison, Toni. NY: Penguin, 1970.

Pages 124–125: *Reservation Blues.* Alexie, Sherman. NY: Warner Books, 1996.

Pages 126–127: *In Dubious Battle.* Steinbeck, John. NY: Penguin, 1936.

Pages 129–130: "Every Subject Must Contain within Itself Its Own Dimensions." In *The Story and Its Writer.* Wharton, Edith. 4th ed. Ed. Ann Charters. Boston: Bedford Books, 1995.

Pages 132–133: *Pygmalion.* Shaw, George Bernard. Mineola, NY: Dover, 1994.

Pages 135–137: *Jane Eyre.* Bronte, Charlotte. NY: Norton, 1971.

Pages 138–140: *Trifles.* Glaspell, Susan. 1916.

Pages 142–143: *Frankenstein.* Shelley, Mary. NY: Bantam, 1984.

Pages 143–145: *The Island of Dr. Moreau.* Wells, H.G. NY: Penguin, 1988.

Music

Pages 155–156:
- www.vervemusicgroup.com/history
- www.apassion4jazz.net/jazz_styles.html
- www.pbs.org/jazz

Pages 157–158:
- www.wikipedia.org
- www.anecdotage.com

Pages 162–163: *La Musica Nuevo Mexicana: Religious and Secular Music from the Juan B. Rael Collection,* Library of Congress American Memory. Lamadrid, Enrique. www.memory.loc.gov.

Pages 170–171:
- www.incwell.com
- www.mozartproject.org
- www.members.tripod.com
- *The Grove Concise Dictionary of Music,* Oxford: Oxford University Press, 1988.

Science and Nature

Pages 181–182: Greenpeace. www.greenpeace.org/international_en/
features/details?item%5fid=328552. (Adapted.)

Pages 185–186:
- www.bcdirectories.com/seasonal/dstime
- http://webexhibits.org/daylightsaving/c.html

Pages 188–189:
- *The New Dictionary of Cultural Literacy, Third Edition.* Ed. James
    Trefil, Boston: Houghton Mifflin, 2002.
- *The Columbia Encyclopedia, Sixth Edition.* Farmington Hills:
    Thomson Gale, 2001.
- www.wikipedia.org

Pages 194–197:
- *The Da Vinci Code.* Brown, Dan. NY: Random House, 2003.
- www.evolutionoftruth.com/goldensection/spirals.htm
- David Yarrow. www.championtrees.org/yarrow/phi/phi1.htm

Pages 198–200:
- *What is Ivory?* U.S. Fish and Wildlife Service. www.lab.fws.gov/
    ivory/what_is_ivory.html. (Adapted.)

Sports and Games

Pages 214–215: *My Body the Billboard.* Johnston, Ian.
www.mala.bc.ca/~johnstoi/.

Pages 216–217: *The Cruise of the Snark.* London, Jack. 1911.

Pages 227–228: *Scribner's Magazine.* Volume 19, Issue 4, April, 1896,
Richardson, Rufus B. www.memory.loc.gov.

Pages 231–232: *The Three Cutters.* Marryat, Frederick, 1835.
(Adapted.)

Social Studies

Page 245:
- www.NewScientist.com
- www.CNN.com
- www.bbc.co.uk

Page 247: *Moving Up and Down the Income Ladder,* U.S. Department of
Commerce. Masamura, Wilfred T.

Pages 251–253:
- *Moral Character in the First Three Years of Life.* Institute for
    Communitarian Policy Studies. George Washington
    University. Garfinkle, Norton.
- *On Becoming Baby Wise,* Ezzo, Gary and Bucknam, Robert. 1995.